Powder Metallurgy Design Manual

2nd Edition

Published by
Metal Powder Industries Federation

Powder Metallurgy Design Manual
2nd Edition

Published by Metal Powder Industries Federation

Copyright 1995 Metal Powder Industries Federation, 105 College Road East, Princeton, NJ 08540-6692.
Printed in U.S.A./6-95-5M.

This Design Manual is issued in the public interest solely as an aid to the engineer seeking information about powder metallurgy. Neither MPIF nor any of its federated trade associations assumes or accepts any liability of any kind arising out of the use or nonuse of any of the manufacturing or design procedures, properties, or specifying guidelines mentioned in this manual, nor does MPIF or any of its federated trade associations make any representation or guarantee of any kind with respect to any product manufacturing or design procedure, properties or specifying guidelines mentioned in this manual.

Library of Congress Cataloging-in-Publication Data
Powder metallurgy design manual.
    p. cm.
    Bibliography: p.
    Includes index.
    ISBN 1878954-54-7 : $75.00
    1. Powder metallurgy—Equipment and supplies—Handbooks, manuals, etc. I. Metal Powder Industries Federation.
TS245.P638   1995   95-1126-2
671.3•7•028--dc20   CIP

Table of Contents

1 Introduction 1
   1.1 Overview of Powder Metallurgy Processes 1
   1.2 Conventional Powder Metallurgy (P/M) 3
   1.3 Metal Injection Molding (MIM) 5
   1.4 Powder Forging (P/F) 5
   1.5 Summary 8

2 The Design and Development of P/M, MIM and P/F Components 9
   2.1 Assessing the Component Operating Environment 11
   2.2 Process Selection 13
     2.2.1 Material Cost Drivers 13
     2.2.2 Tooling 14
       2.2.2.1 P/M Tooling 14
       2.2.2.2 MIM Tooling 14
       2.2.2.3 P/F Tooling 14
     2.2.3 Processing 15
       2.2.3.1 P/M Processing Cost Drivers 15
       2.2.3.2 MIM Processing Cost Drivers 16
       2.2.3.3 P/F Processing Cost Drivers 16
     2.2.4 Secondary Operations 17
       2.2.4.1 P/M Secondary Operations 17
       2.2.4.2 MIM Secondary Operations 18
       2.2.4.3 P/F Secondary Operations 19
     2.2.5 Summary 19
       2.2.5.1 Tradeoffs Within P/M, MIM and P/F 19
       2.2.5.2 Tradeoffs Versus Alternate Processes 19
   2.3 Shapes and Forms of P/M, MIM and P/F Components 22
     2.3.1 P/M Shapes and Forms 22
       2.3.1.1 Pressed and Sintered Shapes 23
       2.3.1.2 Shapes and Features Readily Compacted 25
       2.3.1.3 Shapes and Features Made by Pressing and Sintering 36
       2.3.1.4 Shapes and Features Made by Removing and Displacing Metal 38
       2.3.1.5 Fastening and Joining 39
     2.3.2 MIM Shapes and Forms 40
       2.3.2.1 Beginning the Design 40
       2.3.2.2 Design Considerations 41
       2.3.2.3 Design Features 42
       2.3.2.4 Dimensional Precision and Surface Finish 48
     2.3.3 P/F Shapes and Forms 48
   2.4 Designing P/M, MIM and P/F Components to Meet Structural Criteria 52
     2.4.1 Tabulated Mechanical Properties 53
     2.4.2 Other Structural Properties 54
     2.4.3 Temperature Effects 55
   2.5 P/M Bearing Applications 55
     2.5.1 Bearing Alloys 56
     2.5.2 Design of Bearings 58
     2.5.3 Designing Bearings into P/M Components 58

2.6 Surface Treatment..58
    2.6.1 Electroplating 59
    2.6.2 Resin Impregnation 60
    2.6.3 Mechanical Methods of Surface Treatment...60
    2.6.4 Processes Specified for Ferrous Components 61
    2.6.5 Processes Specified for Stainless Steel Components 61
    2.6.6 Processes Specified for Aluminum Components 61
    2.6.7 Other Surface Treatment Processes 62
2.7 Prototyping 62
    2.7.1 Prototyping Options for P/M...62
    2.7.2 Selecting the Proper Prototyping Option for P/M 63
    2.7.3 Prototyping Options for MIM and P/F 64
2.8 Guidelines for Specifying P/M 64

3 Composition and Properties of Alloys and Materials 67
3.1 Powder Characteristics 67
3.2 Powders Used for P/M 70
    3.2.1 Powder Materials for Structural Applications..70
        3.2.1.1 Ferrous 70
        3.2.1.2 Electromagnetic Materials 72
        3.2.1.3 Stainless Steels...73
        3.2.1.4 Copper Base 74
        3.2.1.5 Aluminum...75
    3.2.2 Powder Materials for Bearing Applications..76
        3.2.2.1 Bronze 76
        3.2.2.2 Ferrous 76
3.3 MIM Feedstocks..77
3.4 P/F Materials...77
    3.4.1 P/F Powders..77
    3.4.2 Mechanical Properties of Powder Forgings..79
3.5 Engineering Properties of P/M Alloys 79
    3.5.1 Density and Related Properties..79
    3.5.2 Mechanical Properties 81
    3.5.3 Other Properties...83
    3.5.4 Summary 84

4 Manufacturing Processes...87
4.1 Conventional Powder Metallurgy (P/M) 87
    4.1.1 Compaction..87
        4.1.1.1 Cycle Events 87
        4.1.1.2 Classes of Compaction and Levels of Features..90
    4.1.2 Sintering 93
        4.1.2.1 The Sintering Operation 93
        4.1.2.2 The Sintering Atmosphere 96
        4.1.2.3 Mixed Phase Sintering...96
        4.1.2.4 Infiltration and Assembly 97
    4.1.3 Secondary Operations 98
        4.1.3.1 Restriking 98
        4.1.3.2 Impregnating 99
        4.1.3.3 Heat Treating Steel Components..99

        4.1.3.4 Machining 100

        4.1.3.5 Surface Treatment 101

        4.1.3.6 Welding/Brazing 103

4.2 Metal Injection Molding (MIM) 104

    4.2.1 Feedstock Preparation 105

    4.2.2 Injection Molding 105

    4.2.3 Debinding 106

    4.2.4 Sintering 106

    4.2.5 Secondary Operations 106

    4.2.6 Trends in MIM Processing 107

4.3 Powder Forging (P/F) 107

    4.3.1 Preform Compaction 107

    4.3.2 Preform Sintering 109

    4.3.3 Forging 109

    4.3.4 Effects of Process on Material Properties 111

    4.3.5 Secondary Operations 112

5 Case Studies 113

6 Glossary 130

7 Bibliography 139

8 Index 140

Introduction

Many designers are turning to powder metallurgy processes (P/M processes) as cost effective alternatives to die casting, forging, gravity casting, investment casting, stamping, fine blanking and screw machining. Reduced development time and performance at minimum cost are some of the factors influencing their decisions. P/M processes produce reliable, precision components at production rates of up to several thousand pieces per hour at near net or net shape.

1.1 OVERVIEW OF POWDER METALLURGY PROCESSES

There are five basic manufacturing processes that utilize metal powders. They are conventional powder metallurgy (P/M), which is the dominant sector of the industry, metal injection molding (MIM), powder forging (P/F), hot isostatic pressing (HIP) and cold isostatic pressing (CIP). This manual will focus on the first three, which are most commonly specified. Further information on the other two can be obtained from the Metal Powder Industries Federation.

P/M processes can utilize a variety of alloys, giving the designer a wide range of material properties. This manual focuses on four major alloy groups: iron and alloy steels, stainless steel, aluminum and copper. In addition, P/M and MIM are used to fabricate materials such as tungsten carbide, tungsten and other heavy metal alloys, super alloys, titanium, tool steels and precious metals.

The three P/M processes covered in this manual give the designer latitude to design components in a wide range of sizes, and in configurations that may range from intricate to simple in design. In most cases zero draft can be achieved, eliminating the need to perform finish machining operations or make design compromises. P/M processes are especially well suited for components requiring fine finishes and bearing properties, and features requiring close production tolerances, irregular shapes, long holes, and blind holes. Cams, gears, sprockets, levers, fasteners, structural mountings, bearings, impellers and hydraulic components are some of the applications being made using these advantages.

Creative utilization of P/M processes through component redesign can often eliminate or reduce multiple assembly operations. For example, designs requiring combinations such as a cam and gear, a spur and pinion gear, or an armature and shaft can be redesigned for replacement by a single, one-piece P/M component. One-piece assemblies can also be created by joining two P/M components during sintering, commonly known as sinter bonding (Figure 1-1).

The versatility of P/M processes lend them to a wide range of application areas. Automotive engines, transmissions and chassis assemblies rely on durable, dependable P/M process components. Aerospace, agricultural equipment, small and large appliances, business machines, defense, electrical instruments, lawn and garden equipment, lock and hardware components, medical equipment, off-road machinery, power and hand tools, sporting goods, and writing instruments are just some of the com-

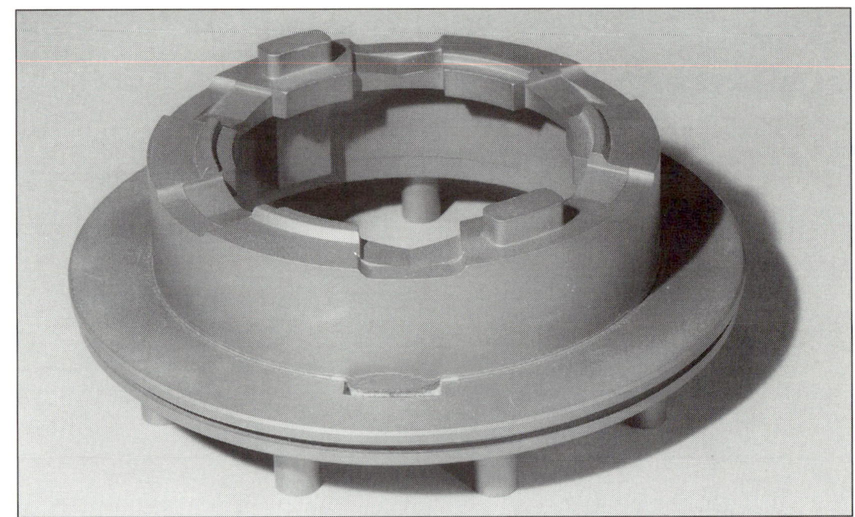

Figure 1-1 This 140 mm (5.5 in.) O.D. by 28 mm (1.12 in.) long counter-balance piston is made in two parts which are assembled as compacted, then simultaneously sintered and brazed together.

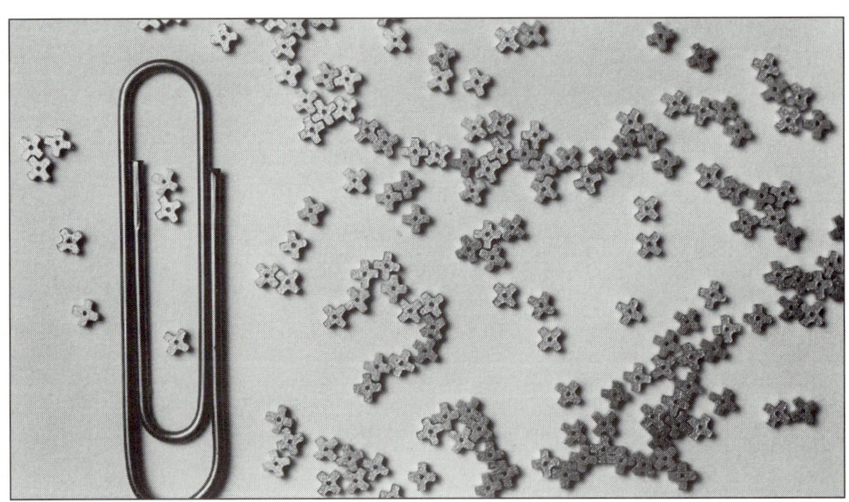

Figure 1-2 These high density pure copper parts, used in a computer printer, have a 0.34 mm (0.0135 in.) I.D. 100,000 parts weigh 1.7 kg (3.7 pounds).

mon application areas where P/M processes are currently being used to advantage.

P/M processes are economical and dependable alternatives because:

- Production and tooling facilities are readily available and lead times are reasonable.
- Raw material sources are stable.
- The process lends itself to be more quickly and readily expanded.

P/M processes are material and energy efficient. The finished product typically utilizes 97% of the metal powder consumed. P/M processes are more energy efficient than casting processes that make similar components.

P/M processed components can be plated to provide decorative, corrosion resistant, or wear resistant surfaces. Individual features requiring additional dimensional precision or increased strength can be coined or repressed. Features that are not practical to form by compaction can be achieved by machining.

Most P/M processed components weigh less than 2.2 kg (five pounds), although P/M compacting presses and furnaces can process components weighing less than a gram (Figure 1-2) to as much as 16 kg (35 pounds).

1.2 CONVENTIONAL POWDER METALLURGY (P/M)

The high purity, custom mixed and alloyed powders used in P/M are fed into a die, compacted into the desired shape, and ejected from the die. The component is then sintered (heated) at a temperature below the melting point of the base material in a controlled atmosphere furnace to form metallurgical bonds between the powder particles. Mixing metal powders in the solid state allows opportunities to engineer material properties unique to P/M. The sequence of operations, with optional post-sintering processes, is shown schematically in Figure 1-3.

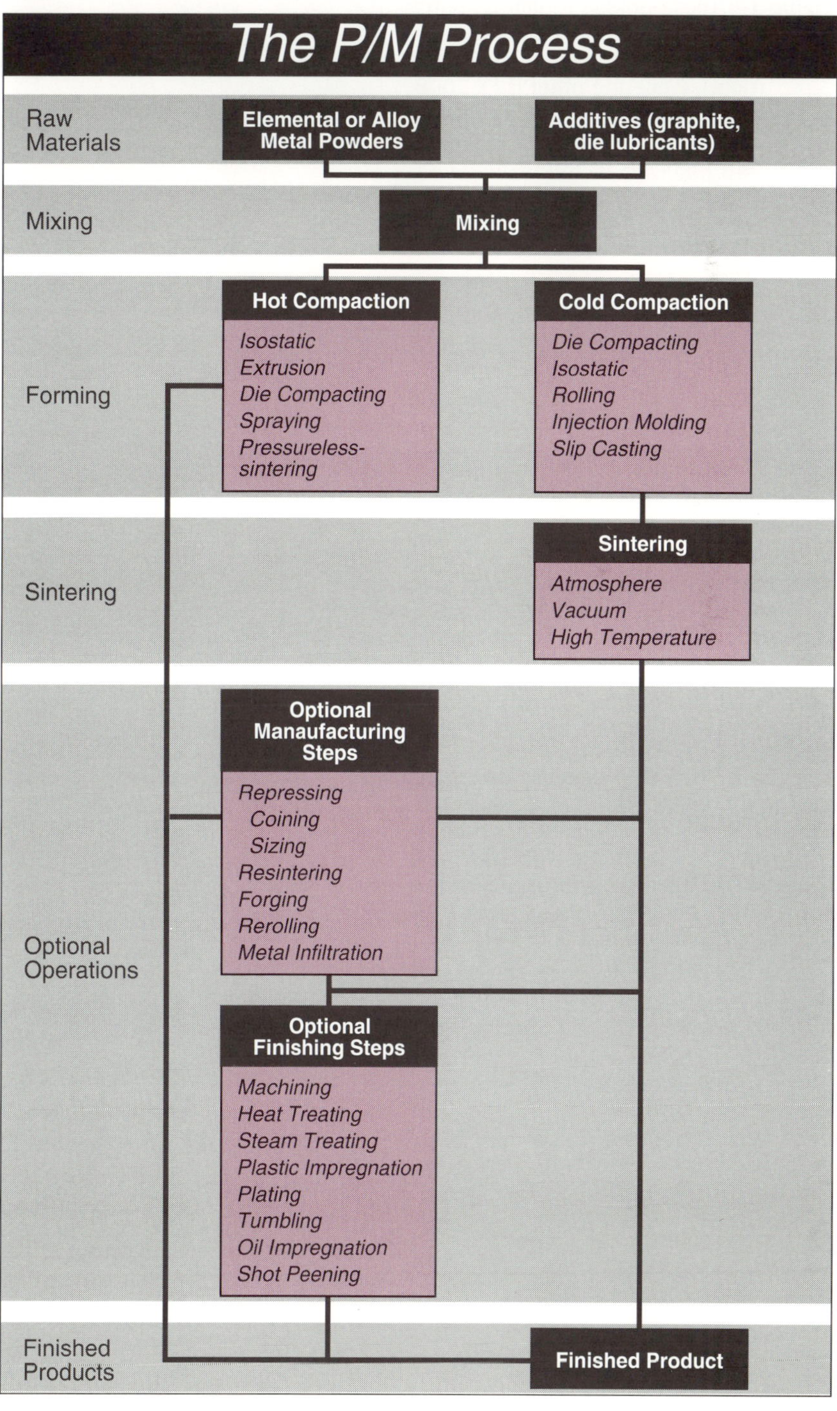

Figure 1-3 The P/M Process.

Figure 1-4 These unique multilevel parts work together in a high security door lock. One is FC-0208 copper alloy steel (right), while the other is FX-2008 copper infiltrated steel (middle). Both are heat treated and the outer part, a key cam (right), is oil impregnated for self lubrication.

Figure 1-5 This 316 stainless steel part, used in forming food products, illustrates the density (porosity) control feature in making P/M parts. The P/M filters are pressed to maintain approximately the desired 30% porosity. Actual size 35 x 19 x 19 mm (1.38 x 0.75 x 0.75 in.).

The P/M structure can have controlled microporosity, which has inherent advantages, such as sound and vibration dampening. A variety of processes are employed that offset disadvantages or utilize microporosity to develop special product properties. Components can be impregnated with oil and solid lubricants to function as self-lubricating bearings, resin impregnated to seal interconnecting microporosity, infiltrated with a lower melting point metal to increase strength and impact energy, and steam treated to increase corrosion resistance and seal microporosity (Figure 1-4). The amount and characteristics of the microporosity can be controlled within limits through powder characteristics, powder composition, the compaction process and sintering process. A common application of controlled porosity is in filters (Figure 1-5).

1.3 METAL INJECTION MOLDING (MIM)

MIM provides the latitude to produce a range of components that are more intricate than those made by conventional P/M. The process begins with the injection of a mixture of metal powder and binder into a mold. Both the powder and binder are specially formulated for the process. The metal injection molding process is very similar to plastic injection molding and high pressure die casting, and it accommodates much the same shapes and features. After molding and debinding, the components are sintered to achieve properties that approach the cast condition. The process is shown schematically in Figure 1-6.

The MIM process is inherently more costly than conventional P/M. It is used to make relatively small, highly complex products that would require extensive finish machining or assembly operations if made by any other metal forming process (Figure 1-7).

1.4 POWDER FORGING (P/F)

P/F processing begins with a preform, which is made in a P/M process that is modified for the application. The preform is converted to its final form in a closed die hot forging operation. Microporosity is virtually eliminated, so that P/F components exhibit significantly higher ductility, impact energy and fatigue strength than P/M. The material content of the preform is held to very close limits so that the forging is essentially free of flash, and close dimensional tolerances can be held. The process is shown schematically in Figure 1-8. P/F is most useful for shapes that can be formed by closed die hot forging where high precision is required (Figure 1-9).

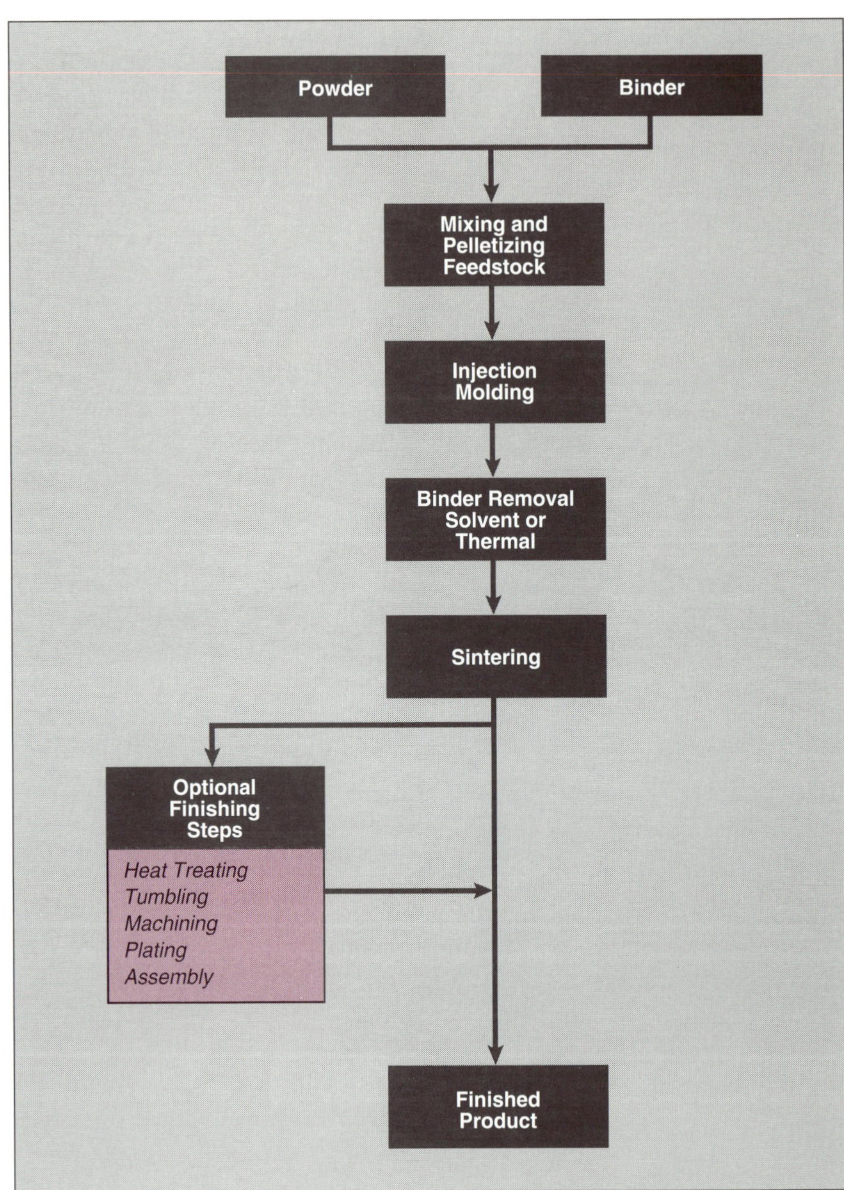

Figure 1-6 The metal injection molding process.

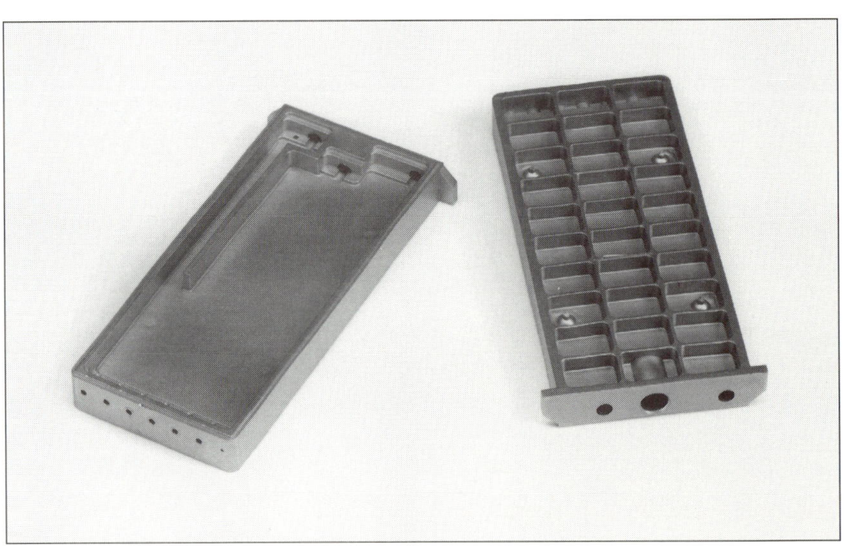

Figure 1-7 Complex MIM microwave module.

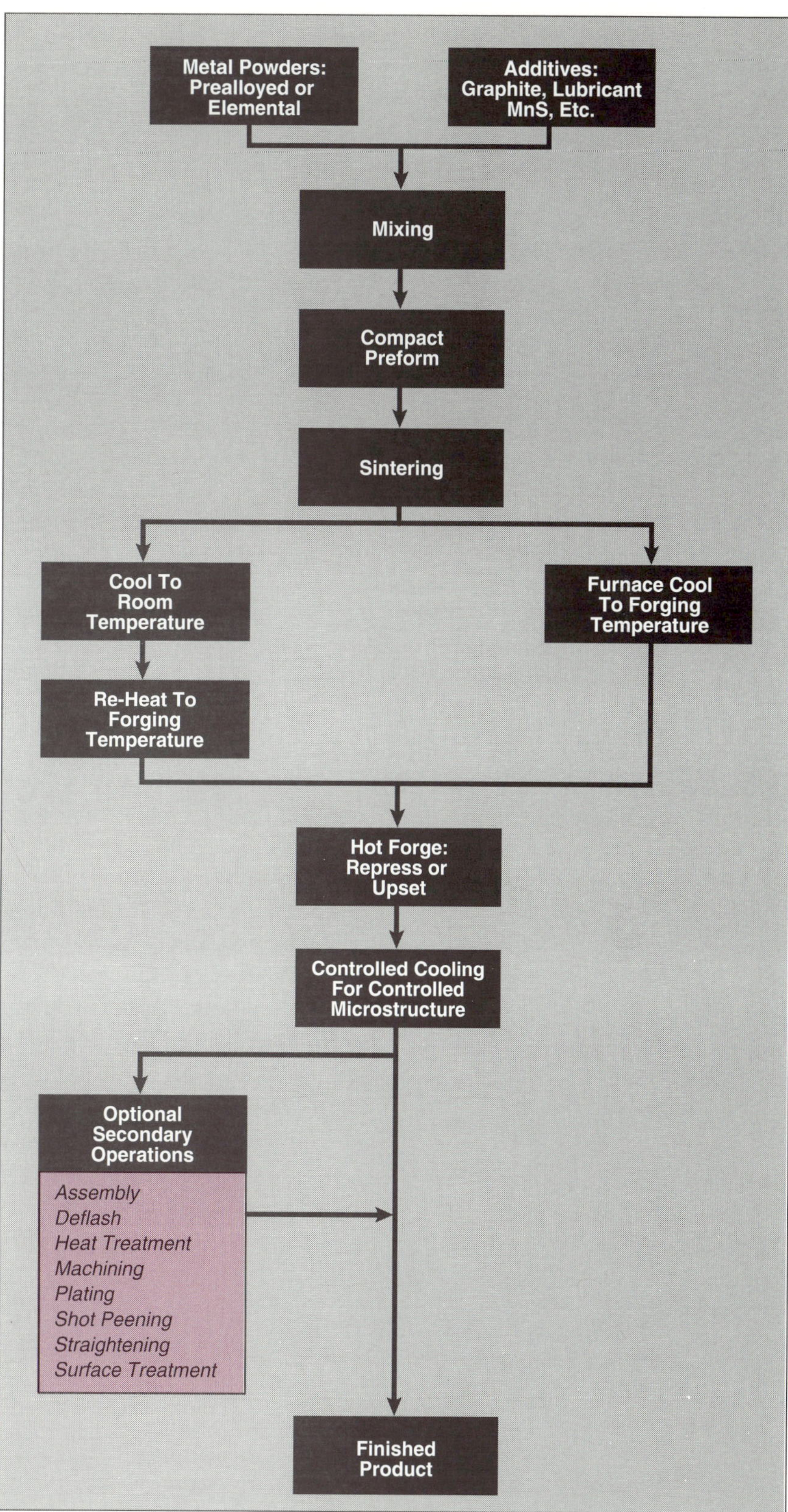

Figure 1-8 Flow diagram for metal powder hot forging process.

Figure 1-9 P/F parts for automotive and industrial applications.

1.5 SUMMARY

There are distinct benefits, which make P/M processes the optimum choice for the designer seeking a material/process that:

- Eliminates or minimizes machining.
- Eliminates or minimizes material losses.
- Maintains close dimensional tolerances.
- Utilizes a wide variety of alloy systems.
- Produces good surface finishes.
- Provides components that may be heat treated for increased strength or enhanced wear resistance.
- Provides part-to-part reproducibility.
- Provides controlled microporosity for self lubrication or filtration.
- Facilitates manufacture of complex or unique shapes that would be impractical or impossible with other metal working processes.
- Suited to moderate-to-high volume component production.
- Offers long term performance reliability in critical applications.

The Design and Development of P/M, MIM and P/F Components

The steps taken to develop a product for the P/M, MIM and P/F processes are similar to those for any other process. Several unique properties and characteristics of the materials and manufacturing processes give the components distinct advantages. To make the best use of these advantages, techniques relevant to the selected material and process must be applied. This is important when a new component is being developed, and especially when an existing component made from another material or process is redesigned for a new process.

The physical form of a component is governed partly by its function. In some cases, form has little to do with function, but is driven instead by the material and process employed. For example:

1. A casting may consist of relatively thin walls reinforced by an intricate system of ribs because those features are most readily cast. A comparable MIM component would have a structure similar to the casting. A comparable P/M component would have somewhat thicker sections with less reliance on ribs, perhaps utilizing lightening holes in non-critical areas, to achieve the same strength at minimum material consumption.

2. A steel stamped or fine blanked component has essentially uniform metal thickness in all features because it is formed from sheet steel. A corresponding P/M, MIM or P/F component can have varying thickness, within limits, according to the requirements of each feature.

Therefore, when considering an alternative process, it is essential to ignore the traditional form of the component under development and focus instead on the function that it is to perform. Form then follows function. For example, if the double gear shown in Figure 2-1 were made in one piece by machining, a relief slot for machining and chip clearance, approximately 5 mm (0.20 in.) wide, would be required where the small diameter gear intersects the face of the large diameter gear. The one piece P/M design shown in the figure has a small boss slightly larger in diame-

Figure 2-1. The P/M process allows the teeth of the smaller gear to be formed full profile to the intersection of the boss with no relief slot required. The helical designs are formed during compacting.

ter than the tooth O.D. The teeth join the boss with no relief required.

The following brief outline will help the designer to determine if a component is a candidate for P/M, MIM or P/F.

1. Creative design begins with a concise, descriptive statement of product objectives. Before attempting to construct such a statement, it will be helpful to review Section 5, Case Studies and Section 2.3, Shapes and Forms. The case studies illustrate the unusual and innovative as well as traditional applications. Section 2.3 gives techniques for reducing the application to the optimum form.

2. As the designer reviews the product constraints, reference should be made to Section 2.2 Process Selection. In this section, P/M, MIM and P/F are compared with each other and with alternate materials and processes, and the types of features that can be made to advantage are identified.

3. The preliminary design phase can be the make-or-break point of the product development program. Therefore, it is essential to optimize the component according to the selected material and process.

4. The efficiency of the product development process is enhanced with early involvement of a qualified manufacturing source, from the preliminary design phase forward. Buyer's Guides are available from the Metal Powder Industries Federation, giving a comprehensive guide to manufacturers of precision engineered components and their capabilities.

   A product engineer who is conversant with tooling and manufacturing processes will be able to interact with the P/M, MIM or P/F engineer. Section 4, Manufacturing Processes, provides the product engineer with the necessary working knowledge.

5. As the design concept is converted to preliminary drawings, Sections 2.4 through 2.6 provide information on topics such as structural considerations, bearing applications and surface treatment.

6. When the preliminary design has been finalized and appears viable, it is usually necessary to fabricate and test prototypes to verify the design. Section 2.7, Prototyping, will guide the product engineer into a strategy that will yield the most product performance information for the least expenditure of time and money. More importantly, the section points out which properties are and which are not correctly assessed in the prototype process.

7. The final design stage incorporates relevant revisions identified as a result of prototype fabrication and testing, therefore fine tuning the design for maximum economy. Section 2.8, Guidelines for Specifying P/M, MIM and P/F, should be consulted at this stage.

8. Production sample procurement and testing normally completes the product development process. At this point a review of Section 2.7, Prototyping, alerts the product engineer to end product properties that may not have been adequately evaluated in the prototyping phase.

2.1 ASSESSING THE COMPONENT OPERATING ENVIRONMENT

A systematic analysis of the component working environment will indicate many application opportunities for P/M that would not otherwise be apparent. The analysis may be conveniently divided into five categories:

1. Applied loads
2. Operating temperatures
3. Corrosion
4. Interfacing components and systems
5. Accidental or unanticipated conditions

Applied Loads

P/M, MIM and P/F are suited for many applications where components are subject to high static and dynamic loads. Ferrous P/M components are routinely made with ultimate strengths several times as high as that achieved by die cast alloys. In ferrous applications, all three processes respond predictably to heat treatment. The very low microporosity of all MIM and P/F components gives them excellent ductility and impact energy. The impact energy of P/M components can be increased significantly by high density, high temperature sintering and metal infiltration. When resistance to abrasion is required, ferrous components can be tailored to the application through engineered alloys and processes. P/M components impregnated with lubricant or special additives offer unique bearing properties that cannot be matched by cast or wrought alloys.

Operating Temperatures

The temperature of the operating environment must be assessed for its affect on material performance. The assessment may indicate that a material can be utilized in environments where the temperatures appear to be above the recommended maximum. Two important criteria for assessing temperatures are cyclic versus steady state and external versus internal.

When temperatures are cyclic, the maximum environmental temperature is not necessarily the temperature experienced by the component. For example, pistons in gasoline engines with melting points as low as 595°C (1100°F) operate satisfactorily when exposed to temperatures in excess of 1650°C (3000°F), because the high temperature exposures are a relatively brief part of each cycle. Conversely, a turbine wheel exposed to gasses at 1205°C (2200°F), requires special high temperature materials because the flow of hot gasses is continuous and the wheel temperature actually reaches the gas temperature.

Steady state external temperatures affect, but do not necessarily govern, the temperature of the component. For example, a pump in an environment where temperatures reach 120°C (250°F), may conduct fluid internally that is at a much lower temperature. Its operating temperature will be between the two, and may be substantially lower than the external environmental temperature.

Ferrous and copper alloys using P/M, MIM and P/F are often chosen when steady state operating temperatures reach several hundred degrees Celsius, or when application environment temperatures vary significantly. Unlike some alternative materials, such as most nonferrous alloys, engineering plastics and composites, the alloys can be engineered so that mechanical properties remain reasonably stable over a temperature range of -40° to 205°C (-40° to 400°F). Further, iron, stainless steel and copper alloys do not exhibit creep or relaxation in this range.

Corrosion

Two types of corrosion must be assessed: atmospheric and galvanic. Atmospheric corrosion is caused by the chemical action of compounds in the environment, primarily water and water soluble pollutants, on the material. Section 2.6 Surface Treatment provides information on a variety of surface treatment systems used with P/M, MIM and P/F components. Copper and stainless steel components rarely require protective surface treatment, and may be specified for their excellent atmospheric corrosion resistance.

Galvanic corrosion is an electrochemical reaction that occurs at the interface of dissimilar metals in the presence of an electrolyte. The tendency for galvanic reaction depends primarily on the difference in electromotive potential of the metals. Less active metals like iron and copper generally exhibit substantially less tendency for galvanic reaction than magnesium, aluminum, zinc and other more active metals. P/M, MIM and P/F may offer an economic advantage by allowing the use of materials that do not need galvanic corrosion protection.

Interfacing Components and Systems

Interfacing components during assembly often present both problems to be resolved and opportunities to gain advantage. When interfacing components are made from different materials galvanic action may occur, as noted above. Different materials may also have significantly different coefficients of thermal expansion that require special attachment techniques to permit the required motion between components.

The optimum method of attachment may depend upon material properties. For example, staking or swaging can be performed only if the material has sufficient ductility. Material properties may make it necessary or advantageous to revise the method of attachment. A material that is subject to creep may require through bolts with nuts to maintain only compression loads, whereas a P/M component operating below the threshold temperature for creep might employ tapped holes to minimize cost.

P/M, MIM and P/F commonly provide opportunities to combine the candidate component with one to which it attaches. The very complex shapes available by MIM make this process especially attractive for parts consolidation. Modifying several existing components into one P/M component usually generates a cost reduction through decreased assembly cost, reduction in hidden costs (usually buried in burden rates), and often increases product reliability.

Accidental or Unanticipated Conditions

The designer must try to foresee any unusual conditions that may occur, and deal judiciously with them in order to ensure product integrity. These conditions may arise when interacting components fail, temperatures exceed design limits, recommended maintenance is not performed, or the product is improperly used.

It is not practical to design each component so that the product will continue to function satisfactorily in the event of unusual conditions, such as accidents. The designer should, however, try to manage the failure to avoid catastrophic consequences, particularly those capable of inflicting personal injury. The engineered mechanical properties of P/M, MIM and P/F alloys can provide the designer with an added safety factor to meet these goals.

Iron, stainless steel, and copper alloys have high melting points and consequently good performance at elevated temperatures. These properties often provide the designer a margin of safety against catastrophic failure when operating temperatures exceed anticipated limits.

2.2 PROCESS SELECTION

The product designer can assess the economic feasibility of a process without quantifying all of the factors that determine the manufactured cost of a product. Working knowledge of manufacturing processes and the associated cost drivers make it possible to assess, with reasonable certainty, the economic feasibility of a component over the expected product life in a given application. Product cost is ultimately determined by price quotations from competent suppliers.

This section assumes a working knowledge of relevant manufacturing processes. Section 4, Manufacturing Processes, describes the manufacturing operations and processes referred to in the following discussion.

Component manufacturing cost can be divided into four categories:
- Material cost
- Tooling
- Processing
- Secondary operations

2.2.1 Material Cost Drivers

Metal powder is higher in cost than an equivalent alloy in ingot or billet form, because of the energy expended in converting metal to powder, and the processing required to prepare the powder for further use. However, P/M, MIM and P/F processes are extremely material efficient.

In P/M, an accurately controlled quantity of powder fills the die cavity and is then compacted to net or near net shape. The process does not require the metal distribution system associated with casting operations, which must be removed and reprocessed. In MIM, the material in the distribution system is re-ground and re-used with no degradation of material properties, and essentially no loss of material. In P/F, the mass of material in the preform is typically controlled to within one-half percent and formed in dies that produce little or no flash. The precision forming capabilities of all three processes can produce intricate features, features with minimal draft, and close dimensional precision which negate most or all finish machining operations with their finish stock requirements.

The end product typically utilizes 97% of the metal powder purchased. Therefore, material cost is closely approximated by the mathematical product of material cost per pound, component volume, and density. Additionally, the high strength and modulus of elasticity developed by many ferrous and copper alloys often require less material to meet required performance than alternate materials and processes.

2.2.2 Tooling

Tooling costs include the initial investment in tools, plus the maintenance and replacement costs. Accurate estimates of tooling costs can only be obtained by working with the appropriate parts producer. Design input from the parts producer can also reduce tool maintenance and replacement costs.

2.2.2.1 P/M Tooling

The initial cost of P/M tooling depends on the level of sophistication, which is directly related to the complexity of the component produced. When a component requires a multiple member tool set and mechanisms to control the motions of the members, tool cost obviously increases. Costs associated with alternate processes also increase with component complexity.

P/M tools produce a relatively large number of components over their useful life. Die materials are selected to withstand the high compaction pressures and abrasive characteristics of the powders. The tools are not subject to the thermal shock and molten metal attack experienced in some metal casting operations.

The coolant systems and heaters used in die casting dies to control temperatures and remove heat are not required for P/M tools, making P/M tools substantially less costly. P/M tools are generally more costly than a single stamping die set, but the relative cost per stamped component depends on the number of stamping dies required.

2.2.2.2 MIM Tooling

MIM molds are very similar to those used for plastic injection molding and die casting. Therefore their costs are very high compared with tools used for P/M, P/F, fine blanking, stamping and sand mold processes. MIM tooling is usually justified when production quantities are sufficiently high to amortize tooling cost, typically in excess of 20,000 pieces per year.

2.2.2.3 P/F Tooling

P/F requires tools for both compacting and forging processes. Where components are to be forged by hot repressing, the compacting tools impart all of the features of the forging. The cost of the compacting tools, which is driven by complexity, is therefore similar to that of an equivalent component made by P/M. When features are formed by plastic deformation of the preform, the preform tools will be simpler in shape and less costly to produce. The forging tools, which impart the final shape, must have all of the features of the component, and their cost is driven by the complexity of the component.

Maintenance and replacement costs are incurred in both the preform and forging tools. The primary drivers for the forging dies are the applied pressures and the amount of plastic flow. Since plastic flow is not uniform, neither is wear. Wear brings about dimensional changes; therefore requirements for very close dimensional precision may dictate increased maintenance and more frequent replacement of tools. In a design that requires a high amount of plastic flow (upset forging), tool maintenance and replacement costs can become a major contributor to the total product cost.

2.2.3 Processing

Processing costs are driven by factors such as labor, equipment investment, processing rates, energy, and atmosphere consumption. These factors vary among P/M, MIM and P/F due to the differences in the processes.

2.2.3.1 P/M Processing Cost Drivers

P/M processing costs include compacting and sintering operations.

Compacting

Compacting cost is driven by the cost of operating P/M presses, which increases with press size. Capital investment increases and production rates decrease as press size increases. Cycle times are typically faster than metal casting machines because no time is required for solidification and cooling.

The press size required for a P/M component is the mathematical product of maximum cross section area in a plane perpendicular to the direction of press motion, and the compaction pressure, which is generally expressed in tons per square inch (tsi). Compaction pressure, a function of the alloy being pressed and the desired density, may be approximated as follows.

Powder and Density	Compaction Pressure	
	MPa	tsi
Ferrous and stainless steel parts		
medium density	400-550	30-40
high density	550-700	40-50
Copper and aluminum parts	250-550	18-40
high density		

P/M presses generally require less floor space than die casting machines, stamping presses, and plastic molding machines for the following reasons:
- P/M presses operate with a vertical stroke, similar to stamping presses, in an orientation that occupies less floor space than a comparable die casting machine or plastic injection molding machine. In addition, P/M does not require a trimming operation.
- P/M components require only one press unless coining or sizing are required. Stamped components often require a line consisting of several presses to perform consecutive operations.
- P/M presses run at substantially faster cycle times than die casting and plastic molding machines, requiring less equipment. They run slower than stamping presses, but the relative press cost is affected by the number of presses required and the degree of automation, as well as press speeds.

Sintering

Sintering costs consist of the costs of equipment, energy, labor and the cost of atmosphere gasses required to maintain the desired conditions in the sintering furnace. Sintering costs vary among P/M facilities due to factors such as scheduling, utilization, and local energy costs. When infiltration is performed (usually infiltration of ferrous compacts with copper base alloys), additional cost is incurred in compacting and placing the slug of infiltrating metal.

Sintering is a solid state process that does not require heating the base metal to the melting point, nor does it require the relatively high heat of fusion. Because of this, less energy is required than for melting and in-house scrap remelting required with casting operations.

2.2.3.2 MIM Processing Cost Drivers

MIM processing costs include injection molding, debinding and sintering operations.

Injection Molding

Injection molding gives the process the capability to produce very complex and intricate shapes, within current size limitations. The cost drivers are qualitatively the same as for plastic injection molding and die casting, but the quantifying factors are somewhat different. Cost is driven by machine hourly operating cost and molding cycle time. Factors that increase machine size increase injection molding cost, and those that reduce cycle time reduce cost.

In general, as component complexity increases, processing and tooling cost can be expected to increase. Requirements for side pulls tend to increase machine size, since these components must be built into the mold and usually increase its size. The operation of these components may add to the molding cycle time. Stepped parting lines increase the costs of producing and maintaining the tools.

Heat must be extracted from the mold to solidify the binder, so features that facilitate heat transfer reduce cycle time. For example, thin walls reinforced by ribs and fins will solidify more rapidly than relatively thick, unreinforced features. Lightening holes, which are effective in reducing material consumption in P/M designs, do not significantly improve heat transfer.

Debinding and Sintering

Debinding is the process in which most of the binder is removed, by the use of either heat or solvents. During sintering, the remainder of the binder is removed. Sintering is similar in nature to P/M, but may require longer cycles, which imply higher costs. Sintering cost cannot be generalized because new technology is rapidly coming on stream reducing the time and cost.

2.2.3.3 P/F Processing Cost Drivers

P/F processing costs include compacting and sintering the preform and repressing or forging.

Compacting and Sintering the Preform

P/F preforms are compacted in a process similar to P/M. The primary concern is to produce a form with the correct mass and very uniform density to facilitate forging. Sintering is likewise performed in a process similar to P/M. Process variables are usually different than P/M, to achieve different properties, but they are generally controlled to similar limits.

Repressing or Forging

Repressing or forging costs are driven by the size and production rate of the forging press, and process controls. Process variables are closely controlled to provide the correct magnitudes and directions of displacement in all areas of the preform.

2.2.4 Secondary Operations

Secondary operations are those that are normally performed after the processing operations discussed previously are completed.

2.2.4.1 P/M Secondary Operations

All secondary operations that can be performed on wrought and cast components can be performed on P/M. In some cases there is a difference in cost, arising from the microporosity inherent in the P/M structure. The following secondary operations are commonly specified.

- Cold forming
    Sizing
    Coining
- Pore treatment
    Infiltration*
    Impregnation
- Heat treatment
- Metal removal
    Turning
    Milling
    Drilling and tapping
    Grinding
- Deburring
- Finishing
    Burnishing
    Blackening or bluing
    Steam treating
    Plating
    Shot peening
    Glass beading
- Welding and brazing

*Normally performed during sintering

P/M components are coined and sized for the same purposes as wrought and cast components. P/M component microporosity facilitates sizing and coining processes somewhat because it allows metal to be displaced more easily.

Pore treatment processes, which cannot be performed on wrought or cast components, utilize the inherent microporosity of P/M. Infiltration uses a metal with a lower melting point than the P/M component to produce a composite structure with properties that cannot be achieved in wrought or cast materials. The P/M interconnected pores can be impregnated with materials that solidify to achieve pressure tightness, or with a lubricant to develop bearing properties superior to wrought or cast parts.

P/M alloys respond to heat treating in a manner similar to cast and wrought alloys. The P/M process also allows custom material formulations which, combined with heat treatment, develop properties not

Figure 2.2-1 The P/M steel truck door lock striker in the foreground replaced the malleable iron casting in the background. The P/M striker consists of two parts, a cap and body, which are projection welded into one unit. The P/M design eliminated surface finish and flatness problems inherent in the casting.

achievable by other processes. Some quenching media should be avoided, such as carburizing salts, brine, or water, which can become entrapped within the structure and cause corrosion.

Metal removal operations are not always required on P/M components because of their inherent precision forming capabilities. Where they are required, materials to improve machinability can be added to the powder mix. Machining feeds and speeds for higher density components (above 92% dense) are similar to those for wrought metals. Lower density components require feed and speed adjustments. Some operations, particularly grinding, tend to reduce or close surface microporosity.

Finishing procedures for P/M components are similar to those used for wrought and cast parts. Some process modifications are required, particularly for low density components, to prevent processing solutions from becoming entrapped in the pores. P/M microporosity enhances burnishing, blackening and steam treatment operations.

Welding and brazing can sometimes be avoided on P/M components by assembling during sintering. Ferrous components can be welded if required using conventional processes such as TIG, MIG, electron beam, resistance, projection and friction (Figure 2.2-1). The effectiveness of the weld depends on component density and composition. Brazing requires filler metals designed for P/M to prevent excessive penetration of the brazing alloy.

2.2.4.2 MIM Secondary Operations

The major advantage of MIM is its ability to reduce significantly, or eliminate, secondary operations, particularly machining and assembly. The number of such operations that are eliminated may give MIM a cost advantage over other processes. Where finish machining operations are required, parameters such as tool materials and configuration are essentially the same as those for wrought or cast products with equivalent chemical composition. Surface treatment processes likewise are similar to equivalent wrought and cast alloys.

2.2.4.3 P/F Secondary Operations

The high precision capability of P/F generally negates many or all of the finish machining operations that would be required with conventional high precision forging. The process produces components with very low levels of microporosity, which is not interconnected. Where finish machining operations are required, parameters such as tool materials and configuration are essentially the same as those for wrought or cast products with equivalent microstructure and hardness. Surface treatment processes are likewise similar to equivalent wrought and cast alloys.

2.2.5 Summary

The following summary gives guidelines for the suitability of P/M, MIM and P/F compared with alternate materials and processes.

2.2.5.1 Tradeoffs within P/M, MIM and P/F

When product design objectives can be met with P/M, it is the choice over MIM and P/F. Tooling and production costs are lower, and precision is generally higher. When complexity increases, particularly for smaller components produced in large quantities, the elimination of assembly and machining operations drives the choice toward MIM. MIM is currently being used for components weighing up to approximately 100 grams (3.5 ounces). P/F is often chosen where precision components with high mechanical properties, such as tensile strength, impact energy and fatigue strength, are required.

The choice of process is often governed by the shape of the component. P/M is best suited for shapes that can be compacted in, and ejected from, compacting tools. MIM shapes are similar to those produced by plastic injection molding in the size range noted above. P/F shapes are similar to those produced by impression die hot forging.

2.2.5.2 Tradeoffs of P/M, MIM and P/F Versus Alternate Processes

The choice of process may be driven by material requirements, geometric complexity, component size or cost. When addressing cost, it is essential to assess the entire life cycle cost including the costs of manufacturing, processing in-house (engineered) scrap, assembling, finish machining, surface treatment, shipping, servicing and retiring at the end of the consumer life cycle.

Die Casting

The choice between die casting and P/M is often a matter of product size or material requirements rather than economics. The most common die casting materials are alloys of aluminum, magnesium and zinc. Copper alloys are also die cast on a limited basis, and are not included in this discussion. Alloys of iron and stainless steel, which have higher melting points, are not die cast.

Die castings can be made in much larger sizes than P/M, MIM and P/F and in equally small sizes. P/M is the choice when material requirements prevail, such as:

- Very high strength. Some ferrous alloys develop tensile strengths more than three times that of the highest strength die casting alloys.
- High wear resistance and bearing applications, which are satisfied by impregnated ferrous and copper alloys.
- High operating temperatures, which are satisfied by copper and ferrous alloys.

- Corrosion resistance, which is achieved by copper alloys and stainless steel.

In applications where temperatures do not exceed 65°C (150°F), and strength requirements are moderate, zinc die casting may be an alternative to ferrous P/M. The two processes are comparable with respect to dimensional precision and machining requirements. Tooling and processing costs usually favor P/M, especially in low and medium production volumes. Die casting allows greater complexity than P/M, and is equivalent to MIM with superior dimensional precision.

Aluminum P/M is sometimes an alternative to aluminum die casting. Aluminum die castings can not generally receive a clear or color anodize because of the silicon levels required for castability, whereas aluminum P/M components with restricted silicon can be anodized. Die casting can achieve greater complexity than P/M, and is generally as economical at high production volumes.

Sheet Metal Forming

When a component can be made as a single stamping from one die, stamping is invariably more feasible and economical. When a progressive die or several dies are required, the tooling cost and machine operating cost are significantly increased and P/M may be competitive. When an assembly of several stampings is required, the costs of fixturing and welding, added to component and tooling costs, makes P/M more favorable. When the complexity exceeds the capability of P/M, MIM may be an alternate. Relative costs for tooling and processing, as noted above, depend on the number of dies and presses required.

Relative material cost should be carefully assessed, because stampings invariably consume more material than the end product requires, sometimes substantially more. Rolled sheet metal used as the raw material is uniform in width and thickness, and does not always conform to the required blank configurations, thereby generating engineered scrap. Stamping also requires peripheral material for clamping the sheet metal in the die. Therefore, significant amounts of metal are removed from the perimeter, as well as from interior openings. This material is trimmed and scrapped at a small fraction of the original purchase price. P/M, as noted above, offers very high material utilization.

Fine blanking, which runs at a slower cycle than conventional stamping and higher equipment cost, is very economical for components with level 1 complexity. The component thickness is essentially constant as it is with conventional stamping. Fine blanking does not have the capability to produce shapes higher than level 2 complexity. (See Section 4.1.1.2 for a discussion of levels of complexity.)

Foundry casting

In this comparison, foundry casting includes conventional sand mold and evaporative pattern (lost foam) processes. The range of alloys for each process are similar so that tensile properties of foundry castings are generally equivalent to P/M and MIM. P/F offers higher mechanical properties, particularly impact, than foundry casting. Process choice frequently depends on size. Foundry castings can be made in sizes that exceed the capabilities of P/M, MIM and P/F.

Where both foundry casting and P/M are viable, foundry castings usually incur lower tooling and material costs but higher production costs. P/M offers greater precision, and may eliminate most or all of the finish machining operations required for castings. P/M usually is the choice when production quantities are high and the tooling cost can be spread over a large number of components. The break-even point is lowered when P/M precision forming capability eliminates finish machining operations. MIM may be an economical alternative to foundry castings for small, intricate components made to high precision.

Investment casting

Investment casting is a high precision process employing a range of alloys similar to P/M, MIM and P/F. When configurations are required that cannot be formed by P/M, or that require extensive machining when made by P/M, investment casting may be advantageous. For small, complex components, MIM is often an economic alternative to investment casting. Tooling cost for investment casting is lower than P/M, MIM or P/F, but production costs are higher. Investment casting is generally competitive at very low production volumes. The break-even point depends on factors such as dimensional precision and the number of secondary operations required. When high dynamic properties are required, P/F may be the process of choice.

Computerized numerical control and automatic screw machine processes

CNC and automatic screw machine processes incur the lowest tooling cost of any production method. The machines are highly automated, and are not labor intensive. Machine processing costs vary according to the number and extent of machining operations. The raw materials are usually bar, tube, plate or billets, and the processes are marked by very low material utilization. Machining from bar stock may utilize less than 50% of the purchased stock, while CNC machining from billet may utilize even less. In general, as the amount of metal to be removed increases, P/M processes become more favorable. The machining versus P/M, MIM and P/F decision may also depend on production quantities; the machining advantage decreases as production volumes increase.

The low material utilization of machining processes is less of a disadvantage for brass than for ferrous components. Cutting feeds and speeds for brass are substantially higher, tool maintenance is lower, and brass chips bring a relatively high scrap price.

The choice between screw machine and P/M may be governed more by component configuration than by economic factors. Screw machine operations are most advantageous when the component is symmetrical with a central axis with features such as cylinders, grooves, lands and threads. P/M can readily produce essentially any shape that can be ejected from the tools.

MIM can make essentially any shape that can be made by machining, in a comparable alloy, and with comparable properties. The choice between machining and P/F may be driven by requirements for material properties, particularly impact energy. The economic break-even point for MIM or P/F versus machining is driven by the same factors as for P/M versus machining.

Impression die forging

The choice between P/M and impression die forging is often determined by product size. Forging can produce components of much larger size than P/M, but it is not used for very small components. In the size range where either may be used, P/M offers greater dimensional precision than precision forging, but lower strength and dynamic properties, such as impact energy and fatigue strength. Recent improvements in P/M processing technology, such as high temperature sintering, have increased the impact energy of P/M enough to make it viable in some cases where moderate to high dynamic properties are required.

MIM is rarely an alternative to impression die forging. It is most advantageous for smaller, very complex shapes that are not generally produced by forging.

The choice between impression die forging and P/F may be determined by component size and complexity. Impression die hot forging can currently produce components of much larger size and a higher degree of complexity. P/F is advantageous when there is a favorable combination of shape complexity, dimensional precision, weight control, machinability or mechanical properties. Relative processing and tooling costs depend on the number of forming operations required by conventional forging. P/F requires one preform tool and one forging tool. Conventional forging usually employs two or more forging tools, a trim die, and sometimes one or several roll forming operations. The reduction or virtual elimination of flash with P/F reduces or eliminates trimming operations and simplifies parting line clean-up. Mechanical properties are equivalent, but a direct comparison cannot be made because the processes use somewhat different alloys, and may respond differently to heat treatment. P/F offers greater dimensional precision, particularly after heat treatment, which sometimes eliminates finish machining operations.

2.3 SHAPES AND FORMS OF P/M, MIM AND P/F COMPONENTS

The shapes that can be produced by the three processes are determined by the factors noted in Section 2.2.5.1, Tradeoffs of P/M, MIM and P/F Processes. For further information, refer to a discussion of the manufacturing process in Section 4 Manufacturing Processes.

2.3.1 P/M Shapes and Forms

P/M components can be shaped to meet a broad span of sizes and complexities while maintaining the basic advantages of low cost, high quality and high performance. Tool contours and relative movements determine the allowable component shapes. Shapes that permit ejection from the tools can be readily made; however, the design of local features must sometimes be modified to insure adequate tool strength. Shapes that cannot be made by pressing only are developed by:

Metal removal, typically machining operations

Metal redistribution, typically coining and sizing

Metal addition-fastening and joining to other components.

Information in Section 4 Manufacturing Processes will help in understanding the following discussion.

2.3.1.1 Pressed and Sintered Shapes

Component Size

Although there is no known theoretical limit, the maximum practical size for a pressed and sintered component is governed chiefly by powder characteristics, component density and available press capacity. The majority of conventionally pressed P/M components range in projected area (perpendicular to the direction of pressing) from 3.9 to 16,000 mm^2 (0.006 to 25 in^2.) and in length (direction of pressing) from 0.8 to 150 mm (0.03 to 6 in.). The practical length limit is closer to 75 mm (3 in.). For a given metal powder, the lower the required density the larger the component that can be produced on a particular press.

As length increases, it becomes more difficult to maintain consistent density through the component. Component length may also be limited by press capacity or tooling requirements imposed by component design; powder filling difficulties occur at thin wall sections, and core rod length may exceed a practical limit. Questions of length limitations are best resolved by consultation with a knowledgeable member of the Powder Metallurgy Parts Association (PMPA).[1]

Dimensional variations

The tolerances that can be held in the P/M process generally depend on orientation of the dimension relative to the direction of pressing, size, component complexity, type of material, process variations and secondary operations.

Nine factors should be considered:

1. **Orientation.** Radial dimensions (perpendicular to pressing) are primarily controlled by the dimensions of the tools. Axial dimensions (parallel to the direction of pressing) are controlled by the motions of the press and the accuracy of filling the die, and generally require greater tolerances.

2. **Component size.** Dimensional tolerances are proportional to size in the radial direction; therefore, large components usually require greater tolerances than small ones.

3. **Component complexity.** The complexities of the tooling and press operation increase with component complexity. Relative movement of multiple tool members and the press members that control them contribute to component tolerances.

4. **Run-out.** The run-out tolerance on components with a hole such as gears, is affected not only by die accuracy, but also by the required running clearances between moving members of the tools. On components with hubs or multiple hubs for which a number of concentric tool members are used, the initial run-out tolerance will be larger than for single level components.

5. **Flatness.** Flatness is a function of component thickness and surface area. Thin components usually tend to distort more than thick ones during sintering or heat treatment. Components with com-

---

1. The PMPA is one of the trade associations that constitute the Metal Powder Industries Federation (MPIF).

plex shapes and unequal cross sections may have density variations that make flatness more difficult to maintain than in those with simple shapes and cross sections. Flatness can be improved by repressing or grinding as finishing operations. Designing the component with pressed bosses or pads to minimize the total surface area is a common practice to improve dimensional accuracy.

6. **Powder formulation.** Materials differ in the extent to which they are affected by variations in sintering time, temperature and atmosphere, and by variations in density and chemistry.

7. **Tool wear.** Tool wear is affected by powder formulation, component density, tool material, and the number of components produced. As tools wear, outside component dimensions become larger and inside dimensions become smaller.

8. **Heat treatment.** Heat treatment of P/M components causes dimensional changes, as it does to wrought materials. Size changes tend to be larger in low density components because the heat-treating atmosphere penetrates deeply and quickly.

9. **Coining and repressing.** Tolerances can be improved by coining or sizing in a repressing operation. Sizing does not necessarily block the pores in medium density and low density components that are to be self-lubricating, but increasing density by coining decreases microporosity and consequently reduces oil-holding capacity.

As with any process, tolerances should be specified no closer than necessary to contain production cost.

Table 2.3-1 lists typical tolerances for radial dimensions up to 12.7 mm (0.500 in.) for eight different materials. Table 2.3-2 lists typical tolerances

Table 2.3-1 Typical Radial Tolerances For Dimensions Up to 0.500 in. (12.7mm)[2]

MATERIAL	AS-SINTERED		AS-SIZED		AS-HEAT-TREATED	
	mm	in.	mm	in.	mm	in.
Brass, Bronze	±0.089	±0.0035	±0.013	±0.0005	N/A	
Bronze	±0.089	±0.0035	±0.013	±0.0005	N/A	
Aluminum	±0.051	±0.002	±0.013	±0.0005	(0.013)	±0.0005
Iron	±0.025	±0.001	±0.013	±0.0005	N/A	
Iron-Copper	±0.025	±0.001	±0.025	±0.001	(0.038)	±0.0015
Copper-Steel	±0.038	±0.0015	±0.025	±0.001	(0.038)	±0.0015
Nickel-Steel	±0.038	±0.0015	±0.025	±0.001	(0.038)	±0.0015
Stainless Steel	±0.025	±0.001	±0.013	±0.0005	N/A	

Axial tolerance: 0.004" (0.102 mm)

Table 2.3-2 Typical Tolerances for Ferrous P/M Components of Uniform Cross Section[3]
Data not recommended for dimensions greater than 2 in. (50 mm)

	Pressed/Sintered				Coined/Sized			
	Practical		Possible		Practical		Possible	
	mm	in.	mm	in.	mm	in.	mm	in.
Length	±0.130	±0.005	±0.075	±0.003	±0.100	±0.004	±0.075	±0.003
Inside Diameter	±0.150	±0.002	±0.025	±0.001	±0.025	±0.001	±0.013	±0.0005
Outside Diameter	±0.150	±0.002	±0.025	±0.001	±0.025	±0.001	±0.013	±0.0005
Concentricity (TIR)	±0.075	±0.003	±0.050	±0.002	±0.050	±0.002	±0.038	±0.0015
Flatness on Ends	±0.050	±0.002	±0.025	±0.001	±0.025	±0.001	±0.025	±0.001
Parallelism of Ends	±0.038	±0.0015	±0.025	±0.001	±0.025	±0.001	±0.025	±0.001

2 Adapted from P/M Design Guidebook (Princeton, New Jersey: Metal Powder Industries Federation, 1984) p 23
3 Erhard Klar, Ed., Powder Metallurgy Applications, Advantages and Limitations (Metals Park, Ohio: American Society for Metals, 1983) p51

for ferrous P/M components of uniform cross section.

2.3.1.2 Shapes and Features Readily Compacted

A shape or feature can be compacted provided that:

1. It can be ejected from the tooling
2. The tools that form it have strength sufficient to withstand repeated applications of high loads required for pressing.

The following guidelines will assist in the design of fifteen types of features that can be made in conventional pressing operations. Consultation with a knowledgeable PMPA member early in the design process will help to optimize the design.

1. **Wall Thickness.** Minimum wall thickness is governed by overall component size and shape. For components of any appreciable length, walls should not be less than 1.52 mm (0.060 in.) thick, but walls as thin as 0.343 mm (0.0135 in.) are pressed successfully in high volume on components 1.0 mm (0.04 in.) in length. Where the ratio of length-to-wall thickness is 8:1 or more, special precautions must be taken to achieve uniform fill, and variations in density are virtually unavoidable. The tooling required for long, thin walls is quite fragile and may have low life expectancy.

2. **Spherical shapes.** Complete spheres cannot normally be made because the punches required to form them would have to be feathered to zero (Figure 2.3-1). Spherical components require straight or flat areas around a major diameter that allows the punch to terminate in a flat. Components that must fit into ball sockets are repressed after sintering to remove the flats. Hemispheres, such as those used in ball joints, can be readily compacted. Spherical depressions up to a hemisphere are also feasible.

3. **Steps.** Simple steps or levels not exceeding 15% of the overall component length and with adequate draft angles can often be formed

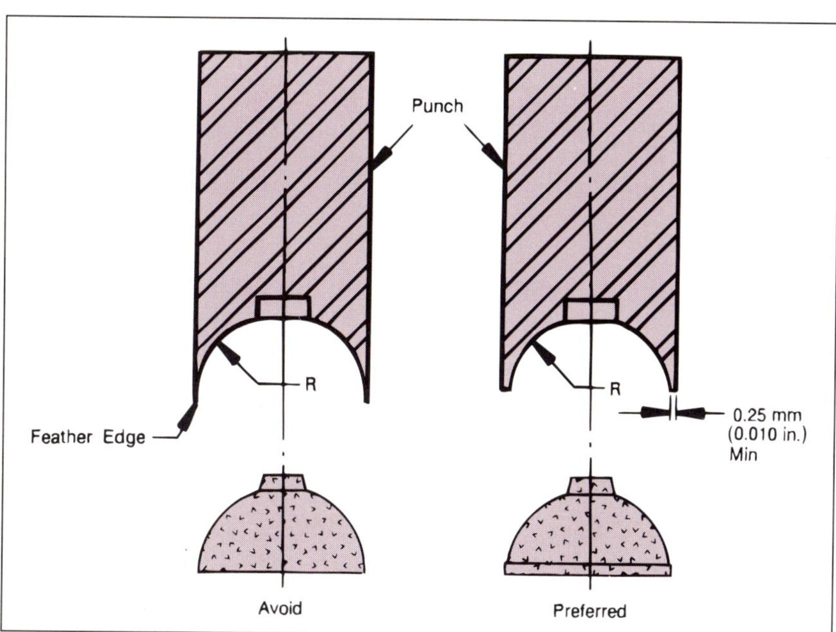

Figure 2.3-1 A flat on the diameter of the sphere allows the punch to terminate in a flat rather than a feather edge.

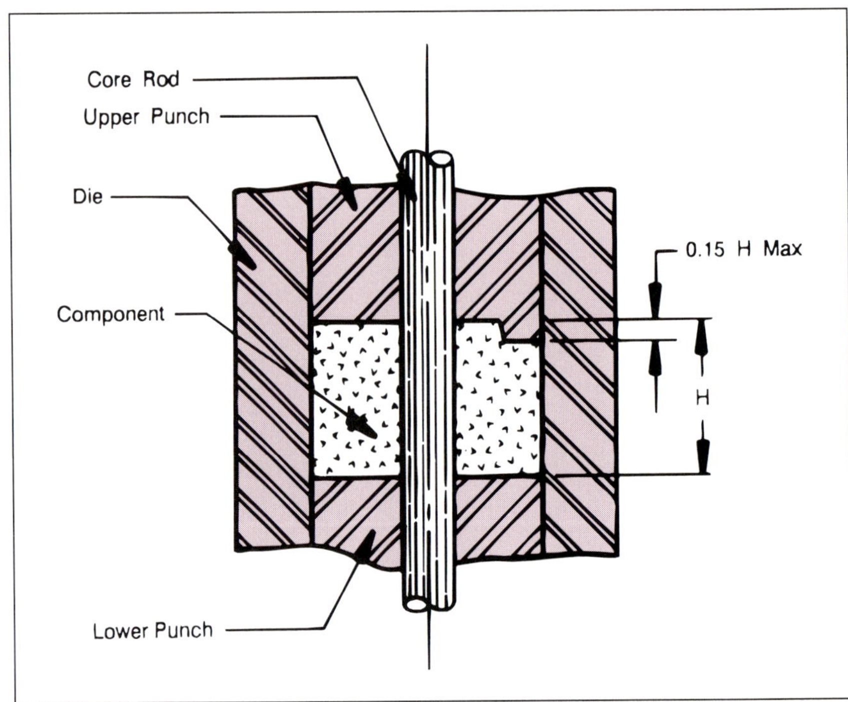

Figure 2.3-2 Simple steps not exceeding 15% of the overall height can be formed in the punches.

in face contours in the punches (Figure 2.3-2). Features such as countersinks and counterbores can be similarly formed (Figures 2.3-10, 2.3-11). There will be variations in density from level to level. However, this approach involves the simplest tooling, resulting in minimum tooling and component cost and closer axial tolerances.

More complex, multiple-motion tooling is required to maintain uniform density throughout components with pronounced variations in levels. Sophisticated compacting equipment, which moves tooling components independently, is used to transfer the metal powder axially before compacting begins, to develop more uniform density. Components with 5 or more levels can be produced with a minimum of density variation between the various levels. Early consultation with knowledgeable PMPA member companies is strongly suggested, particularly during the design of intricate multilevel components, to help keep both the component and tool cost as low as possible.

The minimum preferred width for each step (difference in radial dimension), for multiple-motion tooling should be 1.5 mm (0.060 in.) so that the punches will have adequate strength (Figure 2.3-3). (See previous discussion on wall thickness.)

Projections with large cross sections, formed in short steps, may crack on ejection (Figure 2.3-4) particularly when joined to long (axial direction) features. The projections should be as thick as possible, with radii at the junction.

4. **Alphanumeric characters.** Numbering, lettering, logos, and similar characters can be pressed into surfaces oriented perpendicular to the direction of pressing. They are sometimes added in a secondary coining operation when such operations are otherwise required.

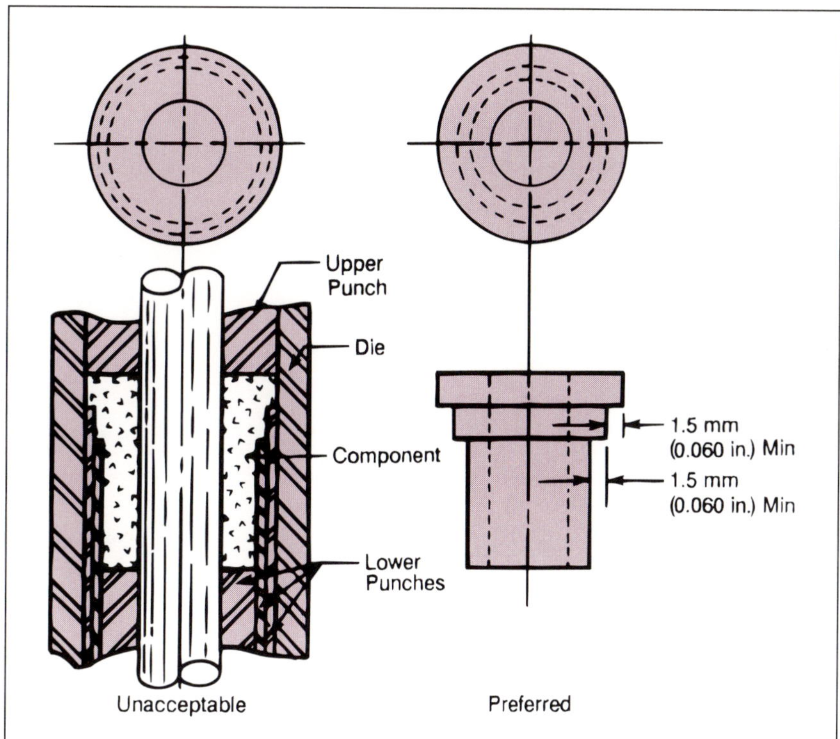

Figure 2.3-3 Multiple steps require a minimum 1.5 mm (0.060 in.) axial difference to allow adequate strength in the punches.

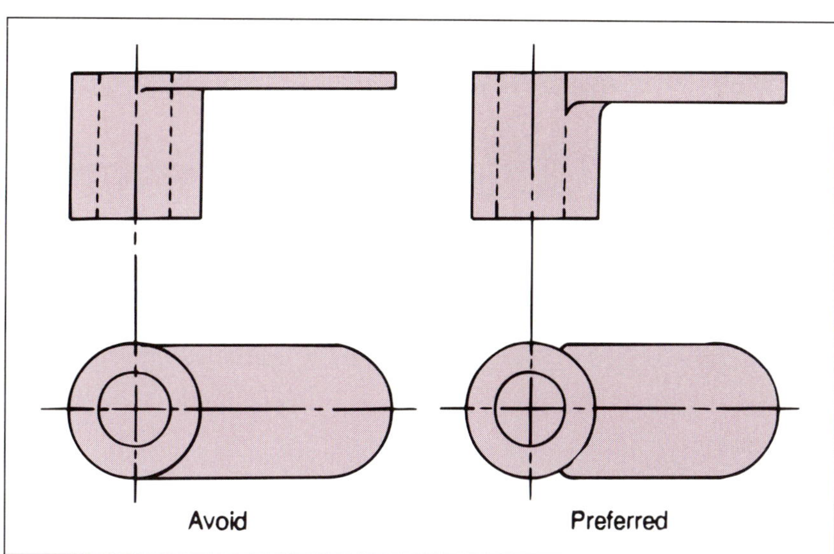

Figure 2.3-4 Large, thin protrusions should be as thick as permissible and be joined to more massive feature with generous radii.

The characters can also be pressed into surfaces that vary from perpendicular, provided that there is sufficient draft on the characters to permit ejection. Either raised or recessed lettering is feasible. However, raised letters are fragile, subject to damage in the green compact, and prevent stacking for sintering.

5. **Tapers and drafts.** Draft is not generally required or desired on straight-through components. While tapered side walls can be produced where required, production speed is reduced to avoid powder wedging between the die taper and lower punch during the fill operation. Tapered sections usually require a short surface parallel

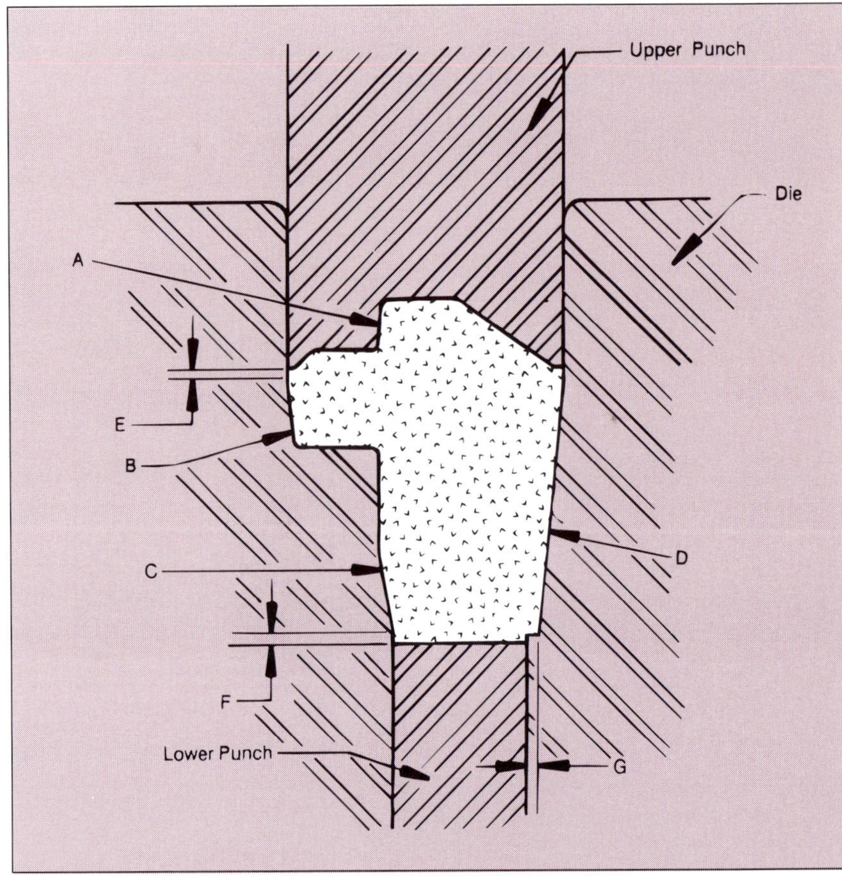

Figure 2.3-5 (Composite drawing to illustrate design principles) A: Step in a punch that forms a shoulder in the component requires up to 12° taper to assist ejection. B: A 2° to 3° taper on an unsupported flange assists ejection. C-D: The maximum feasible taper is 15° when bottom compaction is employed. E: A small parallel surface 0.25 - 0.50 mm (0.010 - 0.020 in.) long, is required where the punch enters the die to prevent the punch from jamming. F-G: Angular forms in the die body can cause powder to be trapped in the acute angle between the punch and die or core rod due to variations in the positions of the punch and die at full compaction. A small surface parallel to the direction of compaction, F, 0.12 - 0.25 mm (0.005 - 0.010 in.) long or a small radial step, G, 0.25 - 0.50 mm (0.010 - 0.020 in.) high can be employed.

to the direction of pressing to prevent the upper punch from running into the taper in the die wall or on the core rod (Figure 2.3-5).

6. **Flanges.** A small flange, step, or overhang can be produced by a shelf or step in the die as shown in Figure 2.3-5. Additional punches and other tooling techniques are required when the amount of overhang becomes too great to permit ejection without breaking the flange. If a stepped die is used, several features are usually required to facilitate ejection and minimize the possibility of chipping the component:
   - Draft around the flange (B)
   - A radius around the bottom edge
   - A radius at the juncture of the flange and body of the component

7. **Holes.** Through holes in the direction of pressing are produced with core rods extending up through the tools, and are readily incorporated in the pressing operation (Figure 2.3-2). Round holes require the least expensive tooling, but other shaped holes such as splines, keys, keyways, D-shapes, squares and hexagons, can be produced with some added tooling cost. Blind holes (Figure 2.3-6), blind steps in holes, and tapered holes, which are usually difficult to machine, are also readily pressed.

The following guidelines apply to the design of blind holes formed by contours in the punch faces:

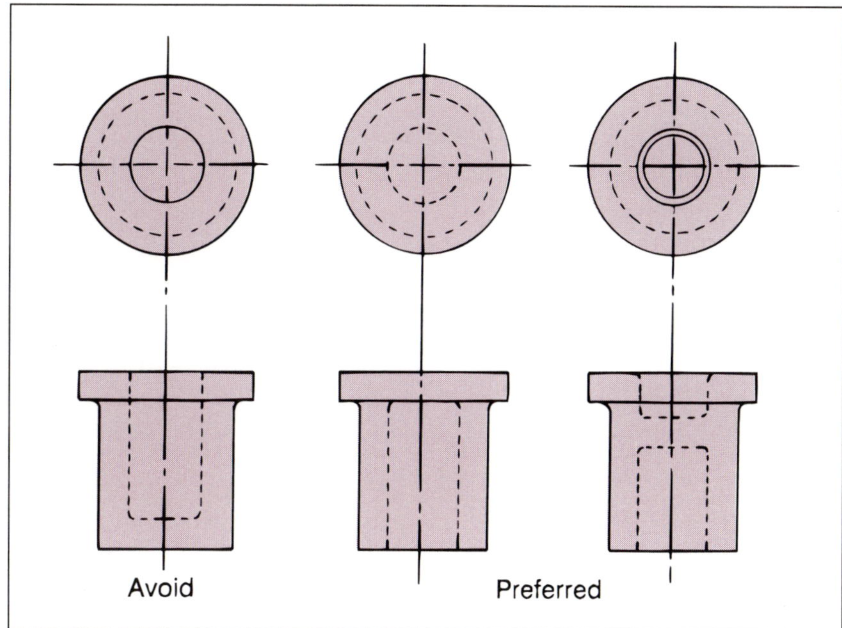

Figure 2.3-6 Avoid blind holes with the blind end opposite a flange.

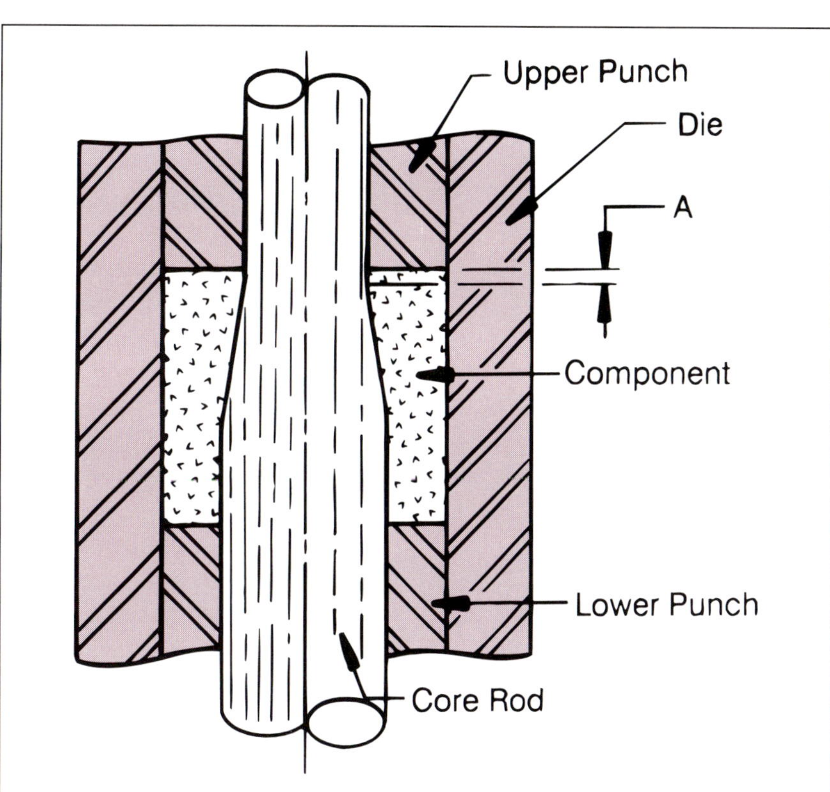

Figure 2.3-7 The tapered hole requires a small parallel surface, A, 0.25 mm (0.010 in.) where the core rod enters the upper punch to prevent the upper punch from wedging on the core taper.

A. The blind end should not be opposite a flange.
B. A short counterbore can be formed on the flanged end (or at either end if there is no flange) provided that:
   • The cross section area of the counterbore does not exceed 20% of the total cross section area
   • The depth does not exceed 25% of the total component length

Figure 2.3-8 Lightening holes are used to reduce material in the lower sproket.

• A minimum taper of 10° to 12° is provided

C. Exceptions to these guidelines, such as the lower counterbore in Figure 2.3-6, can sometimes be made by using a movable core rod or a secondary lower punch.

Tapered holes can be formed, but a short length must be straight to accommodate the punch (Figure 2.3-7). Side holes, or holes not parallel to the direction of pressing, cannot be made in the pressing operation and are generally produced by secondary machining.

Maximum possible diameter of holes in a component may be limited by wall thickness. Minimum diameter depends on the hole depth. The diameter should not be less than 20% of the depth; 2 mm (0.080 in.) is the minimum practical except for very small components pressed by manufacturers who specialize. When the tool length is long relative to its diameter, the core rod that forms the hole can bow during pressing or be pulled out during the ejection cycle. If the component is repressed, the core rod may break due to large lateral or shearing forces.

Lightening holes are frequently used in large components to reduce effective pressing area, reduce weight, and lower pressing force (See Figure 2.3-8). In most cases, these advantages offset added tool cost, especially when the holes are round.

8. **Fillets and radii.** Fillets with generous radii are desirable for most economical design of tools and production. When sharp edges or tight radii are essential to component function, they should be specifically noted on the drawing. For noncritical corners or edges,

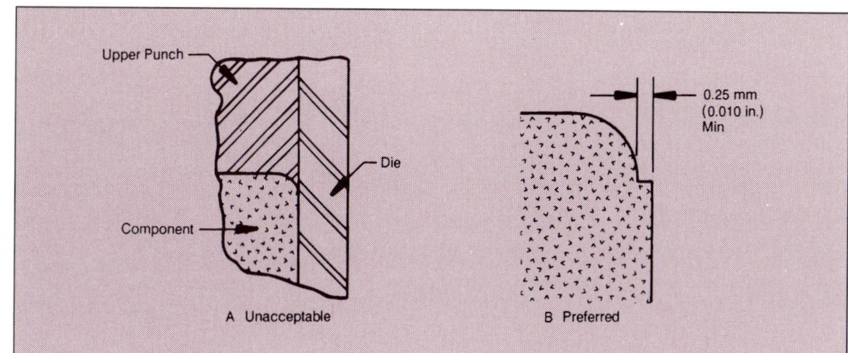

Figure 2.3-9 The radius at the corner (A) cannot be pressed because it would require a feather edge on the punch. The edge may be relieved with a flat (B).

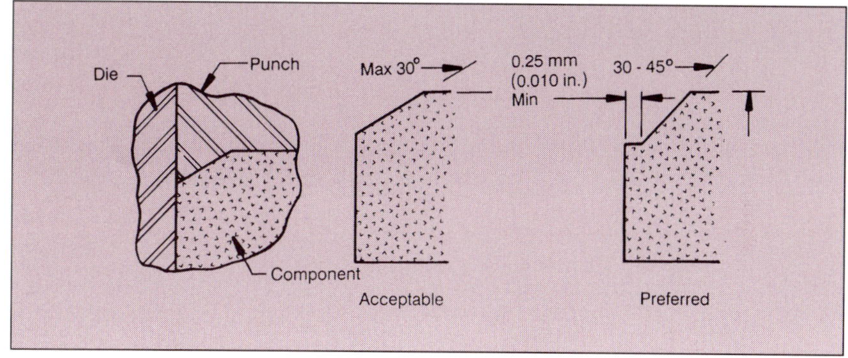

Figure 2.3-10 A chamfer is preferred to a radius at the intersection of tool members. Angles 30° or less from radial provide sufficient tool strength. Angles 30 to 45° should terminate in a flat. Angles greater than 45° should be avoided.

the component producer should be allowed as much flexibility as possible to use fillets or radii on edges that are most suitable for efficient tool construction and component production.

A true radius can not be pressed at the junction of a punch face and a die wall, since it would require that the punch edge or skirt be feathered to zero (Figure 2.3-9). An alternative is shown in the figure. A full radius can be approximated by other processes, such as tumbling. Otherwise, a secondary machining operation is required.

9. **Chamfers and bevels.** Chamfers are preferred rather than radii on component edges to prevent burring. A 45° angle and a 0.25 mm (0.010 in.) minimum flat is common practice to eliminate feather edges (Figure 2.3-10). The preferred chamfer, and the most economical to produce, is a minimum of 60° from axial (maximum 30° from radial) to minimize the chance of breaking the punch protrusion. Although flats are always desirable, their need is less critical as the angle of the chamfer decreases.

When chamfers with an angle less than 60° from axial are required for component function, or when a step would create a problem, the chamfer may be produced by a bevel in the core rod or die. Under some circumstances, bevels as shown in the figure may make tool construction difficult and also decrease production rates, due to fill problems and wedging of powder between the tool members. In such cases, it is often preferable to produce bevels in a repressing operation.

The cost of chamfers can vary considerably, depending on com-

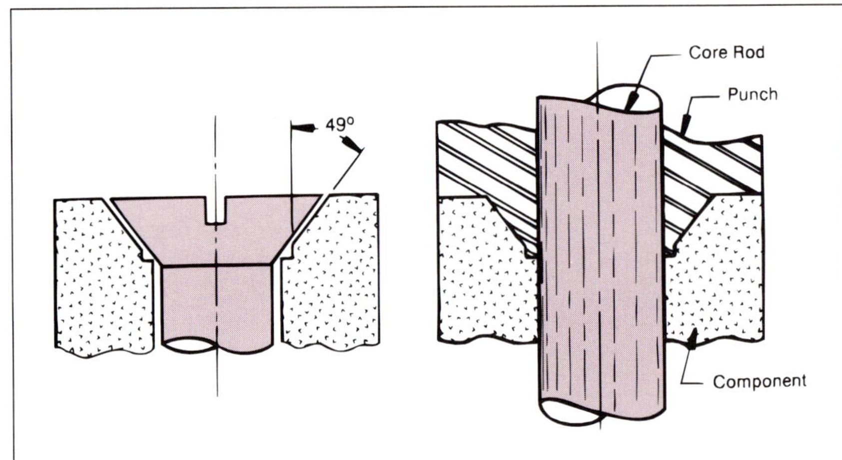

Figure 2.3-11 A standard countersink requires a surface 49° from axial. A small flat at the narrow end provides adequate tool strength without interfering with the function of the feature.

ponent shape and tool design. For example, a chamfer in a plain round component or on two sides of a square component costs relatively little, because the rim or skirt can be easily ground into the punch. On the other hand, chamfers on irregular shapes require more work on the tooling and more tool maintenance, and thus tend to increase overall cost.

10. **Countersinks.** A countersink is a chamfer around a hole for a screw or bolt head. A standard 82° countersink, when produced by the punch, requires a face that is 49° from axial, which is less than the recommended 60°. Therefore, a flat of about 0.25 mm (0.010 in.) is essential to avoid fragile, sharp edges on the punch (Figure 2.3-11).

11. **Bosses.** A boss can be located on either end of a component under the following conditions (Figure 2.3-12):
    1) Length must be small compared to the overall component length-preferably 15% or less.
    2) A round shape is preferable. Other shapes may require relatively costly tool manufacturing methods, such as EDM or an engraving technique to produce the depression in the punch.
    3) Draft angle should be as much as possible, preferably up to 15° per side, to avoid locking the boss into the punch surface. Bosses with sides parallel to the direction of pressing require special punch members to give positive component ejection.

12. **Knurls.** Axial knurls can be made on inside and outside diameters; diamond or angled knurls interfere with ejection. Ground diamond knurls in the punch face reproduce as a waffle structure on the component (the reverse of a true diamond knurl). To reproduce a true diamond knurl, the punch must be engraved or finished by the EDM process.

13. **Hubs.** Hubs that provide for drive or alignment rigidity in gears, sprockets and cams can be readily produced. Features such as keys, D holes, and keyways can be formed in the pressing operation (Figure 2.3-13). A variety of shaped holes in the axial direction are possible. Set screw holes at right angles to the drive axis are produced by secondary drilling and tapping operations (see Figure 2.3-19). A generous radius between the hub and flange is preferred,

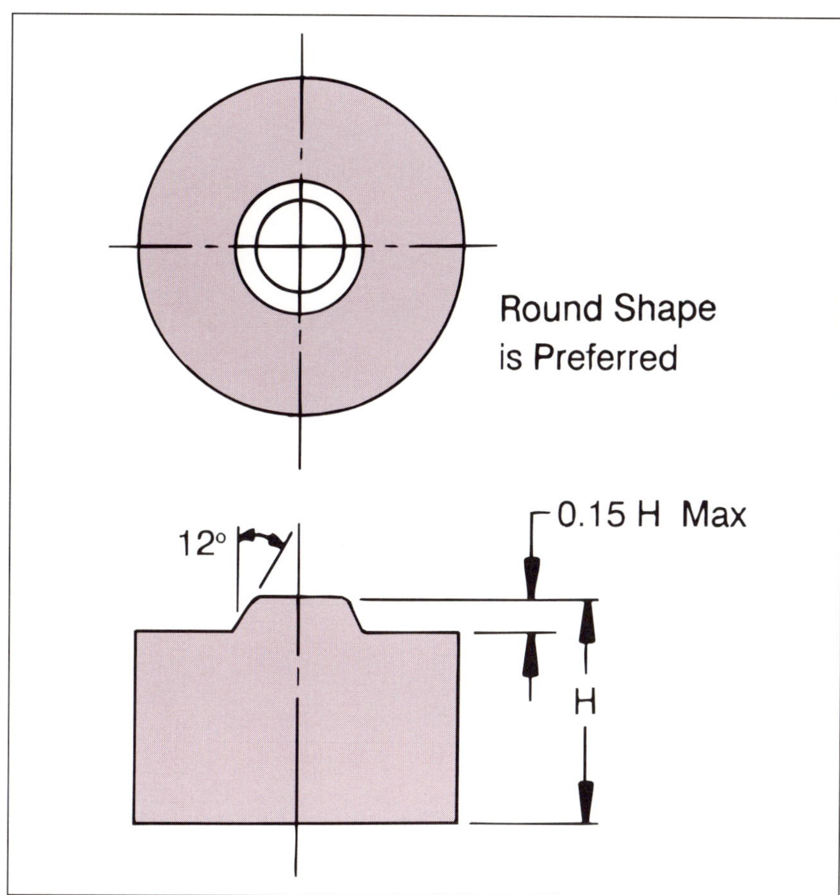

Figure 2.3-12 A boss can be formed in the punch face at either end of the component provided that the guidelines for draft and height are observed.

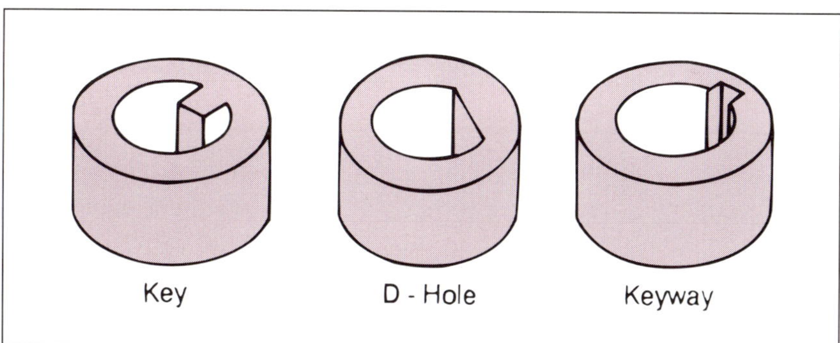

Figure 2.3-13 Keys, D holes and keyways can be formed in hubs in the compaction operation.

as well as maximum permissible material between the outside diameter of the hub and the root diameter of components such as gears and sprockets (Figure 2.3-14). Hubs that contain self-lubricating bearing surfaces can be produced at lower density than other portions of the component to impart additional oil capacity. Relatively shallow hubs can be treated as bosses (see Bosses).

14. **Studs.** Shallow studs with drafted sides are made readily in the tooling in the same manner as bosses. Where no draft is allowed or where length-to-diameter ratio is relatively large, particularly with respect to the overall shape of the component, additional tool members are required for proper compaction and ejection (see Hubs). A small diameter stud or protrusion is not desirable, par-

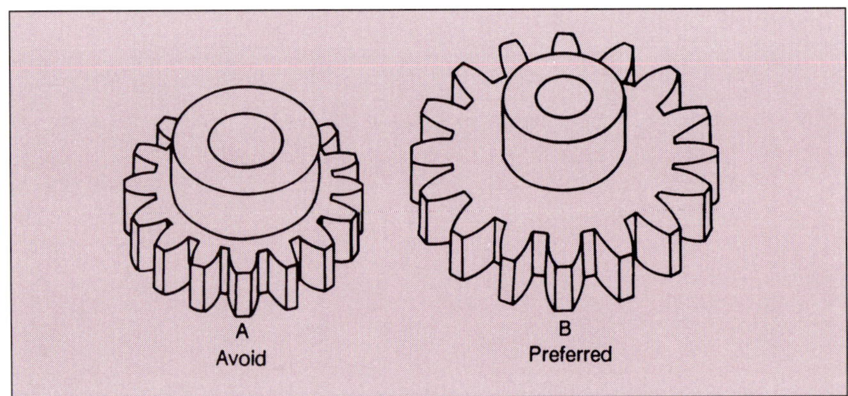

Figure 2.3-14 The design at A allows insufficient material between the hub and roots of the teeth. Design B is preferred.

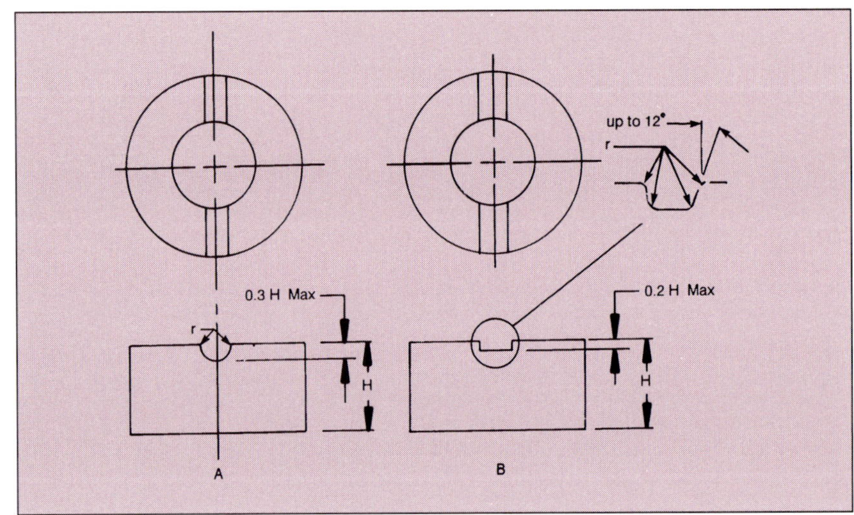

Figure 2.3-15 Semicircular and rectangular grooves can be pressed from a projection on the punch faces into either end of the component provided that the guidelines for draft, height and radius are followed.

ticularly on a comparatively large and heavy component, because it is fragile in the green state. Alternatively, provide a hole in the component into which a pin can be assembled in a secondary operation.

15. **Slots and Grooves.** Grooves can be pressed from projections on the punch into either end of a component within the following limit for low and medium density components (Figure 2.3-15):

   A. Curved or semicircular grooves to a maximum depth of 30% of overall component length.

   B. Rectangular grooves to a maximum depth of 20% of the overall component length, provided that
   - Surfaces parallel to the direction of pressing have up to 12° draft
   - All corners and edges are radiused or chamfered

   Deep, narrow slots and grooves are undesirable because they require similarly shaped tool members, which are subject to breakage. Where possible, they should be redesigned to promote greater tool strength (Figure 2.3-16). Otherwise, the features will require machining.

16. **Undesirable Features.** Features must be avoided that require tools with thin, weak sections or sharp inside corners; cause problems in

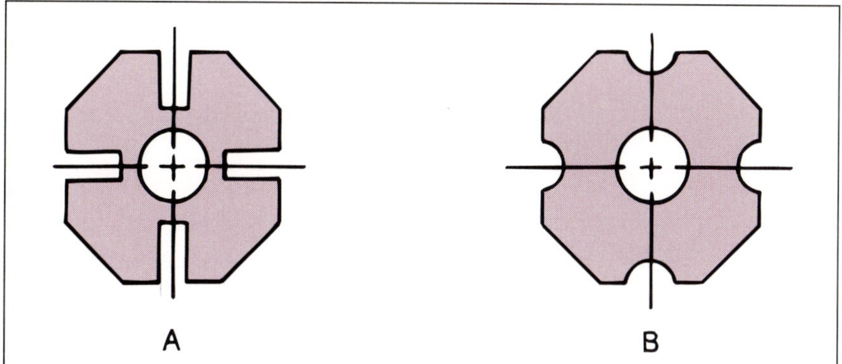

Figure 2.3-16 The design at A requires long, thin tool members and the sharp corners cause problems in powder fill. Design B avoids both problems.

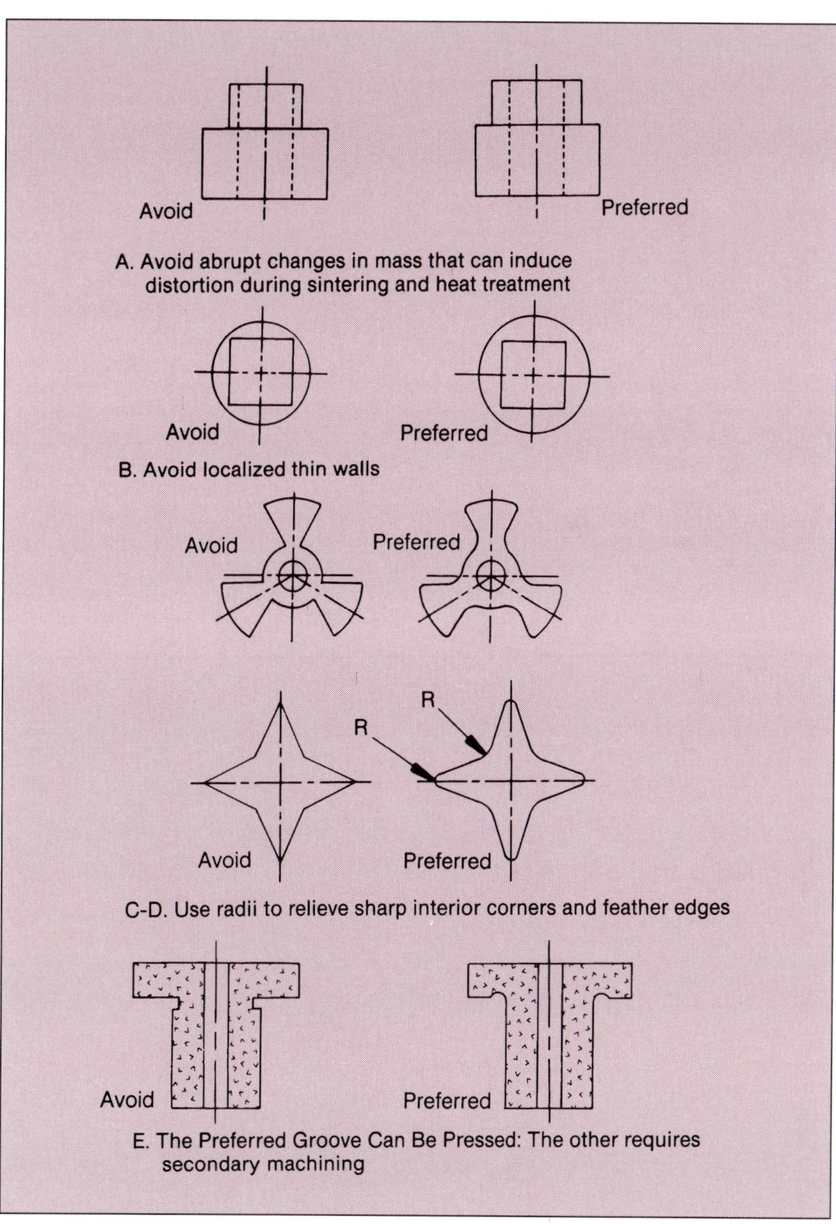

Figure 2.3-17 Features that cause tooling or powder fill problems must be avoided.

powder fill and compaction; or make the component prone to distortion during sintering. Several such features are shown in Figure 2.3-17.

2.3.1.3 *Shapes and Features Made by Pressing and Sintering*

The P/M process can produce complex shapes, such as those shown in Figure 2.3-18, to very close tolerances, provided that component contours allow ejection of the green compact from the die. Shapes that require fragile tools should be carefully considered since P/M tooling is subject to very high stresses.

The P/M process is particularly well-suited to the production of gears for several reasons:

1. Tools can be made to provide consistent component accuracy over long runs.
2. Retention of some microporosity contributes to quiet running gears and allows for self-lubrication.
3. Gears can be made with blind corners, as shown in Figure 2.3-19, eliminating undercut relief that is needed with cut gears and lending extra strength at the blind end.
4. Gears can be combined with other components such as cams, ratchets, and other gears to form a single component.
5. Bevel, miter, helical and other special gear forms are possible. Copper infiltration is sometimes used on bevel or miter gears to increase density in the entire tooth.
6. True involute gears are less difficult and less costly to produce by P/M than by other methods because tooth shape is not a limitation.

Following are specific gear design aids:

1. Location of holes relative to the gear form is affected by running clearances of the various tool members. Therefore, it is more difficult to hold the close runout tolerances obtainable with arbor cut gears. Runout tolerances can be improved by grinding the inside

Figure 2.3-18 The P/M process can produce complex shapes to close tolerances.

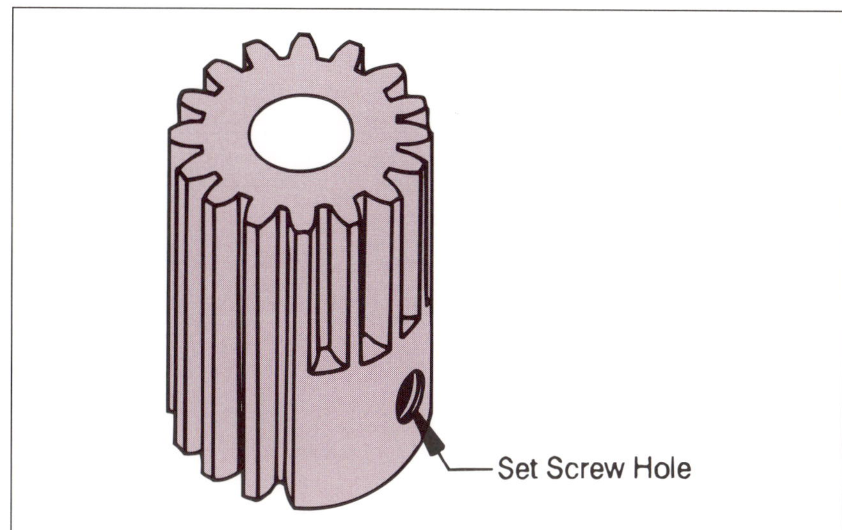

Figure 2.3-19 Gear teeth with blind corners can be pressed. The set screw hole is drilled and tapped.

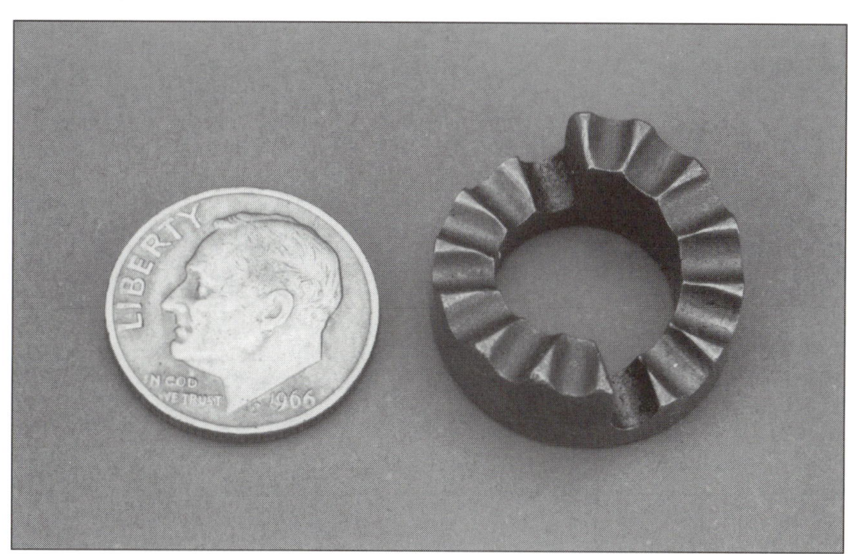

Figure 2.3-20 The stepped cam face contour was formed from features in the punch face.

diameter of the hole true to the pitch diameter of the gear as a finishing operation.

2. Hubs or pinions that increase the number of concentric tool members increase the runout tolerance. These features should be located as far as possible from the root diameter to avoid thin tool sections. Where this is a problem, two-piece assemblies should be considered.

3. As the AGMA class of gear increases, the cost of a P/M gear increase because of additional operations, such as sizing or repressing, boring, honing and grinding.

4. Helical gears can be produced by P/M but require specialized tooling. Helix angles up to 40° are currently being produced.

Ratchets, like gears, are readily produced by the P/M process. Where possible, undercut teeth should be avoided and maximum possible radii allowed at the tip and at the root.

Cams benefit from many of the design advantages of the P/M process, such as excellent surface finish, high component-to-component form consistency and self lubrication. The surface of a self-lubricated P/M cam will often exhibit longer life than a ground cam surface. Radial cam shapes are formed in the die; face cam shapes are formed in the punch faces or as steps in the die (see Figure 2.3-20). Radial and face cam shapes can be combined into one P/M component. Cams can also be combined with gears, ratchets, etc.

2.3.1.4 Shapes and Features Made by Removing and Displacing Metal

When shapes and features are required that prevent the component from being ejected from the die, secondary operations, such as coining and machining, are sometimes employed. This class of features includes:

1. **Undercuts.** A component with an undercut in the flange to allow fit-up to a dead corner, such as shown in Figure 2.3-17E, requires a machining operation. An alternative, which can be pressed, is shown.

2. **Reverse tapers.** A reverse taper (larger on bottom than on top) cannot be ejected from the die, and must be machined.

3. **Annular grooves.** Annular grooves around a component are frequently machined (see Figure 2.3-21). The component can also be made as a two-piece assembly (see Section 2.3.1.5 and Figure 2.3-22).

Figure 2.3-21 The annular groove was machined after sintering.

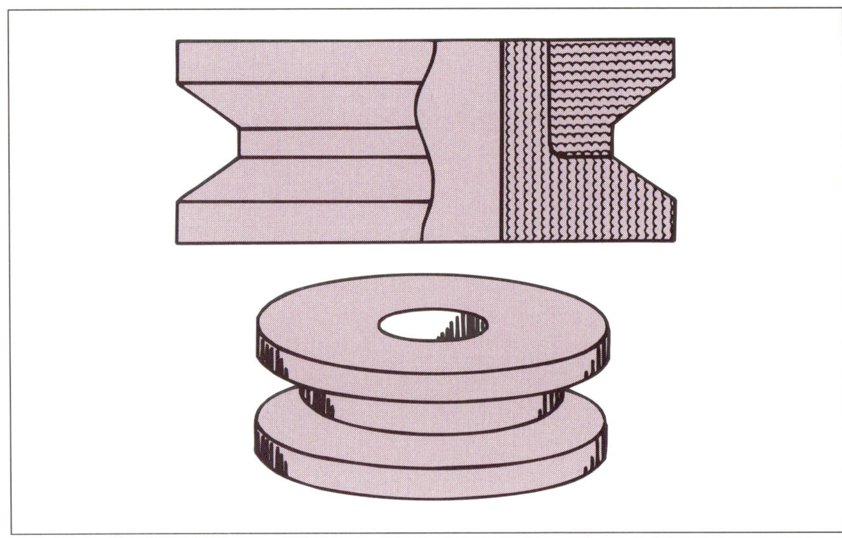

Figure 2.3-22 The pulley is economically made by joining two P/M parts.

4.   **Threads.** Threads in holes and on outside diameters prevent ejection, and are produced by conventional thread cutting operations.

2.3.1.5 Fastening and Joining

Two or more P/M parts often can be joined to form a component that is difficult or impossible to make in conventional pressing operations. Several parts can be joined together to form simple undercuts, eliminate machining operations, or form a component like a V-type pulley (Figure 2.3-22). Two-piece assemblies that can be made in one piece by conventional metal working methods may be made to advantage with P/M as a single component by combining several comparatively simple parts, thus avoiding costly tooling. An assembly can also take advantage of the different properties of various P/M materials, such as a bronze bearing in a ferrous structural component, and a heat-treated component with one that is not.

Conventional methods

P/M components can be joined by conventional methods such as press-fitting, staking, and other methods of upsetting, as well as by brazing, soldering, welding, cementing, riveting and bolting. Brazing often requires infiltration to prevent the component from absorbing excessive quantities of relatively expensive brazing metal. High strength joints can sometimes be achieved by locking mechanically with D-shaped holes, or by key/keyway combinations.

Methods employing differential expansion

Where different materials are involved, components can be joined during the sintering operation, through proper selection of materials, by taking advantage of their different size change characteristics. For example,

Figure 2.3-23 Sinter brazed parts.

outside components can be designed to shrink during sintering, relative to the inside component, to produce a strong bond between the components (see Figure 2.3-22). This method does not always achieve a 100% bond, but the joint strength is sufficient for many applications. Infiltration can increase the strength of the bond as well as of the component itself.

Sinter brazing

Sintering and brazing can be simultaneously performed on iron components such as those shown in Figure 2.3-23 in an operation termed sinter brazing. The parts to be brazed are assembled either green or partially sintered, and a measured slug of braze alloy is placed in the proximity of the joint. Furnace heat melts the braze alloy while sintering occurs, and the braze alloy flows into the parts through capillary action. The process develops a very strong joint with shear strength typically 380 to 415 MPa (55,000 to 60,000 psi).

2.3.2 MIM Shapes and Forms

As noted in Section 2.2.5.1, MIM is currently being used to produce components weighing up to approximately 100 grams (3.5 ounces). Within this size, shapes and forms are almost unlimited, due to the flexibility of the metal injection molding process that produces the basic form. As shapes become more complex, tooling and production costs increase. A variety of alloys are currently in use, such as nickel-iron magnetic alloy; nickel steels; stainless steels such as 316, 410, 440C and 17-4 PH; tungsten carbide; and pure nickel. Alloys of aluminum, magnesium, lead and zinc are not good candidates.

The following design overview is based in part on The Metal Injection Molding Design Guide published by the Metal Injection Molding Association.[2]

2.3.2.1 Beginning the Design

To achieve the optimum balance between complexity and economy, the design should begin with the simplest shape, then depart, as required, one feature at a time, to achieve the desired product property or function. At each step, the cost of increased complexity should be assessed, in terms of tooling and processing costs, and weighed against the anticipated benefits. Production cost is best obtained from MIM parts producers. It is therefore important to consult with them from the beginning of the product design process.

The simplest shape is produced in a mold made of two sections with plane surfaces that meet to seal off the cavity. The cavity is formed by a core that extends from one section and fits into an impression in the other with uniform clearance to produce MIM shapes with uniform wall thickness. The core produces the internal features, and the cavity produces the external features. All features are designed to permit the cavity to release from the solidified form, and the form to be pushed off of the core with ejector or knock-out pins.

In order to develop the desired product function, the designer must usually depart from the simplest shape, adding complexity, and conse-

---

2 For more detailed information, consult the Metal Injection Molding Design Guide, which is available from the Metal Powder Industries Federation.

quently cost, to the tool and process. At each stage, the designer should weigh the value received against the increased cost. Value is measured in terms such as improved product function, reduction in subsequent machining operations, reduction in material consumption, and reduction in parts count. Cost is best assessed by consulting MIM parts producers.

2.3.2.2 Design Considerations

The design must recognize three requirements of the molding process. First, material must be fed into the mold cavity through gates, which are critical to proper fill of the mold cavity, and which may leave a visible mark on the molded form. Second, the mold must open, revealing a visible parting line on the component. Third, the form must be ejected from the mold using ejector or knock-out pins, which may leave visible marks on the molded form.

Parting Line

The location of the parting line is critical to the orientation of component features. It is the beginning point from which draft is established on surfaces that are parallel to the movement of mold members. To the extent possible, all features should be oriented perpendicular to the plane of the parting line to facilitate removing the form from the mold. Features that cannot be thus oriented require additional mold members such as side pulls. Each additional member increases the cost of the tools, and may increase processing cost. Additional mold members may be cost effective when the features they form would otherwise require machining or assembly operations. If component complexity does not require secondary operations or significantly lengthened cycle times, there may be minimal impact on component cost. Complex components, if designed to meet the processing and tolerance requirements of MIM, can deliver the best combination of economy and functional design.

A parting line that can be contained in a single plane is preferred. However, it is often necessary to modify the simple shape in order to mold desirable features. The added complexity increases the cost of fabricating and maintaining the tools. As with side pulls, the complication may be cost effective when the features would otherwise require machining or assembly operations.

Gate Locations

The optimum location of gates is a balance between product and processing requirements. Gates usually leave a mark on the finished MIM part where they are separated from the molded form. Therefore, their existence must be recognized in the design. The optimum location requires input from both the designer and MIM parts producer. In general, gates are located on the parting line, and should be positioned to direct the flow onto a core rod or cavity wall. Where wall thicknesses vary, gates should be located so that material flows from thicker sections to thinner.

Ejector Pins

Ejector or knock-out pins are required to push the molded form off of the mold cores(s). There must be enough pins to free the molded form from the core without distorting it. The pins must be large enough so as not to make too deep an impression on the molded form, which will carry over onto the finished component.

Ejector pin locations are a balance of processing and functional requirements. In order to minimize distortion, they should be located near features that require the highest ejection forces, such as cored holes. They are often located on structural features, such as ribs, which can receive and transmit the ejection forces. The pins leave permanent marks, which become flash collars as wear occurs between the pins and their guides. The pins should bear on recessed features when it is necessary to prevent the flash from protruding above a functional surface. Where possible, the pins should bear on surfaces that will be finish machined or on unused areas, such as runners, which are trimmed off and reprocessed.

Provision for sintering

MIM components are placed on plates or shelves for sintering. A geometry, such as a flat surface, which permits the piece to rest securely with no additional support, is preferred. Long spans, cantilevers or delicate points may require special fixtures or setters to minimize distortion during sintering.

2.3.2.3 Design Features

The design features that can be made by MIM are similar to those made by conventional plastic injection molding or die casting. Dimensions and proportions are different, based on the characteristics of the material being molded. Fundamentally, the mold must fill completely to reproduce all mold details, the feedstock must solidify without damaging the form, and the form must be ejected without distortion, ready for further processing. All features must be designed with these fundamentals in mind.

Draft

Draft, which is illustrated in Figure 2.3-24, is the small angle on surfaces that would otherwise be parallel to the direction of movement of mold members. It is provided to facilitate release and ejection of the molded form. The normal range is 0.5° to 2.0°. The actual amount of draft increases with the complexity of the component or addition of multiple

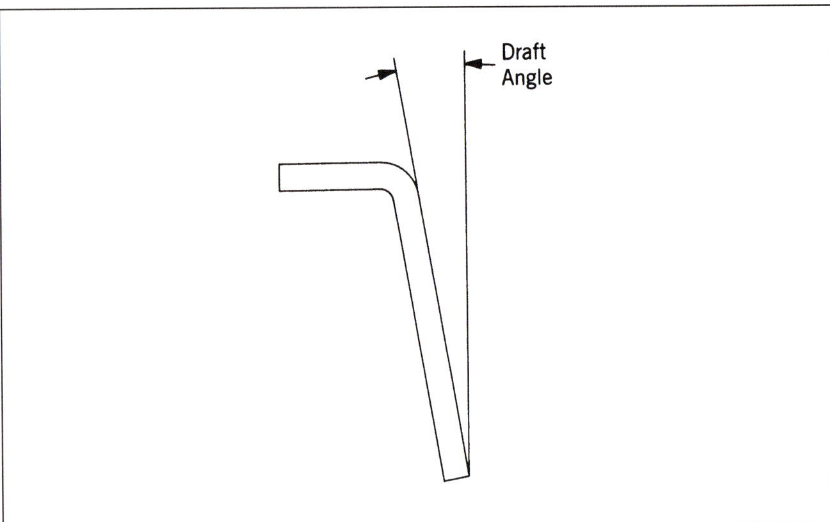

Figure 2.3-24 Illustration of draft angle.

cores. Draft should be uniform on all features to facilitate mold fabrication, and should be near the upper limit, consistent with functional requirements, to facilitate molding.

Wall thickness

Where feasible, walls should be of uniform thickness throughout. Thickness variations lead to distortion, internal stresses, voids, cracking and sink marks. In addition, they cause non-uniform shrinkage, interfering with dimensional and tolerance control. Thicknesses in the range of 1.3 to 6.5 mm (0.05 to 0.25 in.) are preferred, but exceptions in both directions are possible, and MIM suppliers should be consulted for guidance.

Figure 2.3-25 shows several common ways of modifying a form to make wall thickness more uniform. Where wall thicknesses vary, the design should provide gradual transition, as shown in Figure 2.3-26, or a generous radius, as shown in Figure 2.3-27. Cores can be used to advantage to achieve uniform wall thickness as shown in Figure 2.3-28.

Holes

Holes should be cored to reduce material consumption, promote uniform wall thicknesses and reduce or eliminate machining operations. The

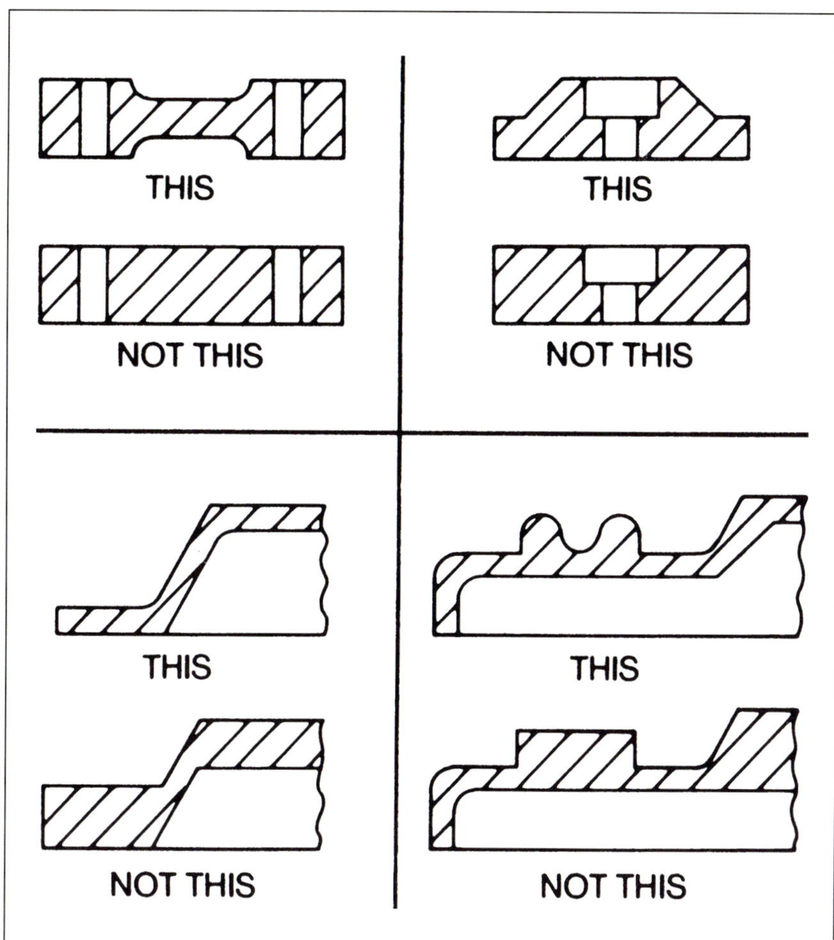

Figure 2.3-25 Uniform walls will minimize or eliminate distortion, stress concentrations, voids, cracking and sink marks, and will improve dimensional control.

preferred direction is parallel to the direction of mold opening, which is perpendicular to the parting plane. Through holes are preferred to blind holes because the core pin is supported on both ends, whereas blind holes use a cantilevered pin. Internally connected holes are possible, but they require special tooling considerations, and they may cause problems in flash removal. They should be perpendicular to each other and, where possible, one should be D-shaped to facilitate core pin sealing as shown in Figure 2.3-29.

Ribs and webs

Ribs and webs are useful for reinforcing relatively thin walls and avoiding thick sections, as shown in Figure 2.3-30. In addition to increasing

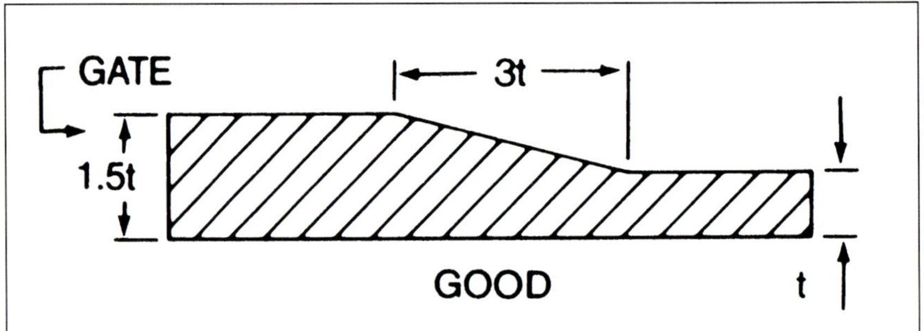

Figure 2.3-26 When wall thicknesses vary, a gradual transition between sections reduces stress concentrations and surface imperfections, and provides for improved flow during filling of the mold.

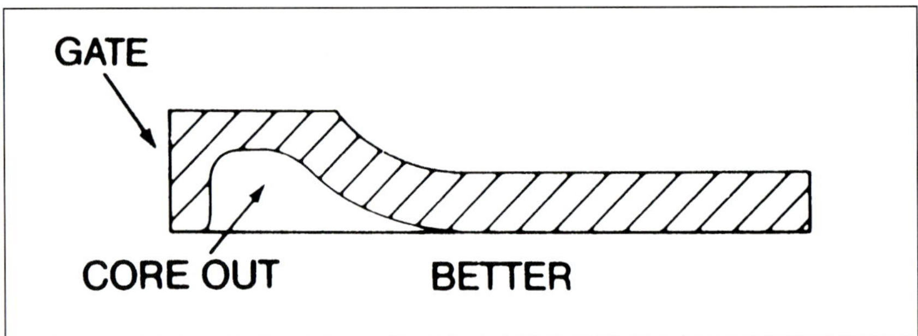

Figure 2.3-27 Removal of excess material by coring can be helpful in attaining a part with uniform wall thickness.

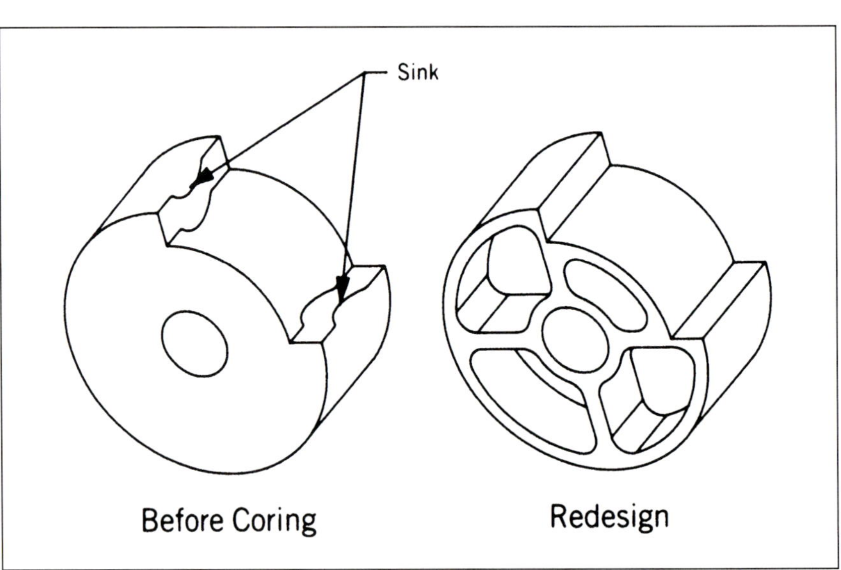

Figure 2.3-28 Cored holes can be used to achieve uniform wall thickness, reducing or eliminating sink marks and distortion

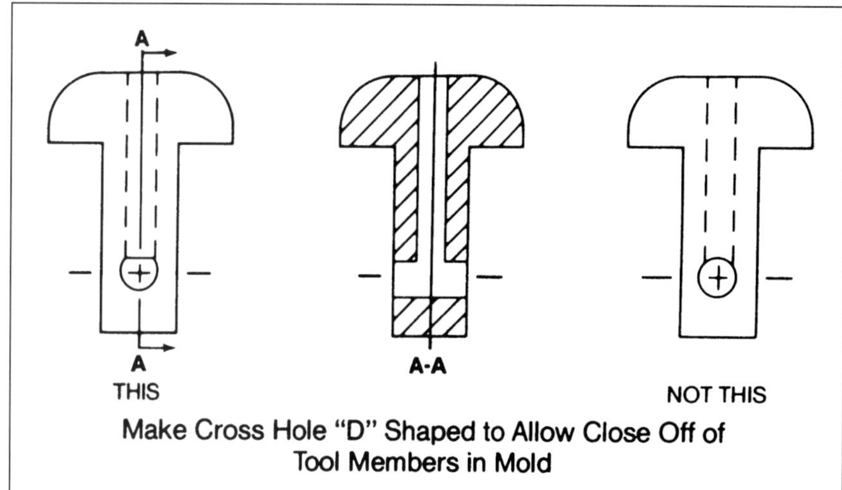

Figure 2.3-29 Intersecting holes should be perpendicular to each other, with one of the holes being "D" shaped to facilitate closing off in the mold.

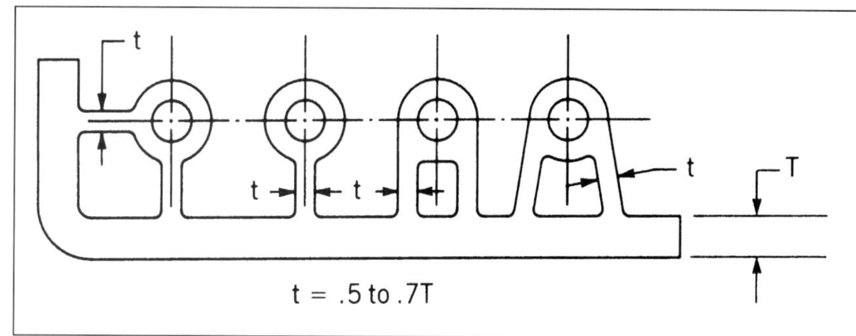

Figure 2.3-30 For thin walled parts, or parts with cross-sections reduced to promote uniformity of walls, reinforcing ribs or webs are an effective way to increase strength and rigidity.

strength and rigidity of a thin wall, they improve material flow and limit distortion. Rib thickness should not exceed that of the adjoining wall. Where structural requirements indicate thicker ribs, multiple ribs should be used instead. Ribs may cause sink marks on the opposite surface of the adjoining wall, warpage, or stress concentrations. Because ribs are formed by removing mold material, it is often advantageous to evaluate performance without ribs, then add them as indicated by performance tests. Recommended proportions for ribs are shown in Figure 2.3-30.

Fillets and radii

Fillets and radii are generally advantageous to product function as they reduce stresses at the intersections of features. They are also advantageous to the molding process, by eliminating sharp corners that can cause cracking or erosion of mold features. Most MIM parts suppliers prefer 0.4 - 0.8 mm (0.015 - 0.030 in.) fillets and radii.

Bosses and studs

Bosses and studs are useful for forming attachment points. The maximum length of a stud or hole should not exceed five times the thickness of the adjoining wall, and the stud diameter should not exceed the thickness of the adjacent wall. A boss with a blind or through hole should have a diameter twice that of the hole, and the wall thickness of the boss should not exceed that of the adjoining surface. These parameters limit hole

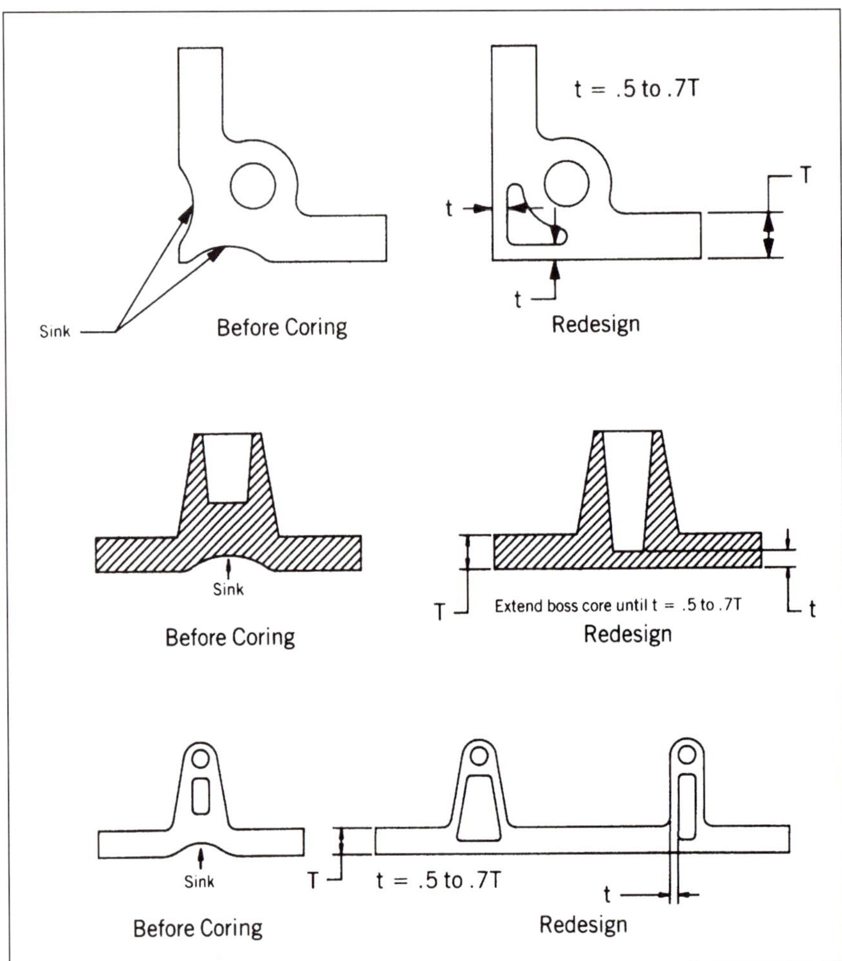

Figure 2.3-31 Bosses and studs are useful for forming attachment points. Wall thickness of associated holes or cored areas should equal .5 to .7 times the thickness of adjacent surfaces.

diameter to twice the thickness of the adjoining wall. Design parameters for holes are illustrated in Figure 2.3-31.

Threads

Both internal and external threads can be formed by MIM, but experience shows that tapping internal threads is usually more precise and cost effective than forming with core rods that must be unscrewed. Unscrewing the core rods adds to tool cost and may impose a slight reduction in production rates. Potential users should consult with MIM producers for specific information on molded-in threads, such as class capabilities and costs.

The optimum location for external threads is on a parting line of mold members, to negate the need for unscrewing the mold members that form them. Adding flats to the threads on the parting surface, as shown in Figure 2.3-32, allows for flash of 0.05 mm (0.002 in.) per side, improves thread quality, and reduces mold maintenance.

Undercuts

External undercuts can be readily formed on a parting line, as shown in Figure 2.3-33, with the same considerations described for external threads. However, where additional mold members are required to pro-

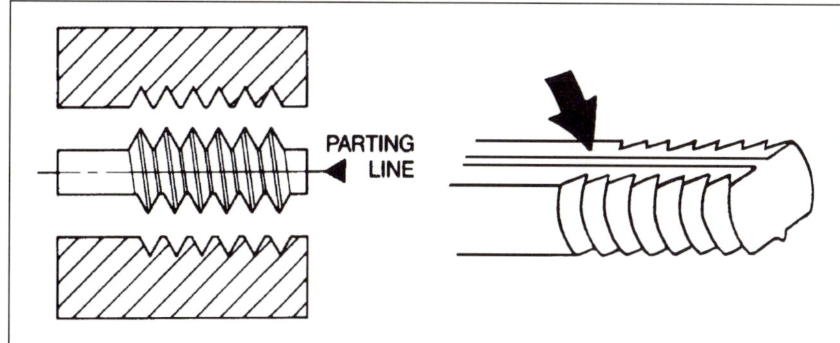

Figure 2.3-32 Adding narrow flats to external threads at the parting line surfaces improves thread tolerances by keeping flash below the root diameter.

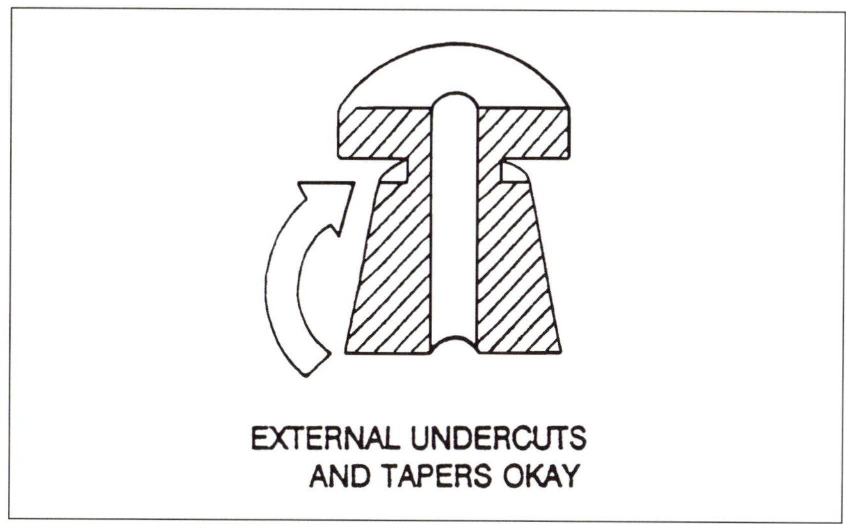

Figure 2.3-33 External undercuts can be readily formed on MIM parts. However, tooling considerations dictate that there be two parting lines 180° apart on the surface of the undercut.

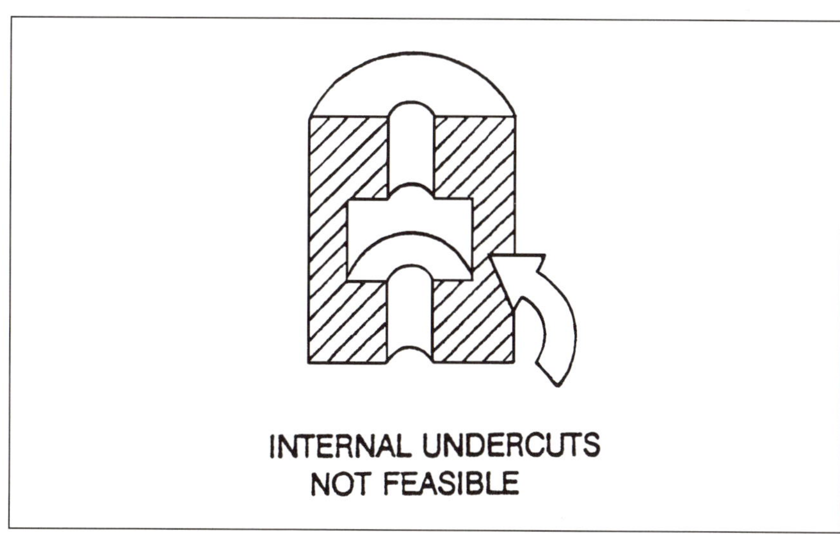

Figure 2.3-34 Internal undercuts or recesses are not feasible in MIM components and should be avoided.

duce them, tool costs are increased and production rates are decreased. When the component can be redesigned to eliminate the undercuts and attendant mold members, tooling and production costs are reduced. Internal undercuts, such as the one shown in Figure 2.3-34, can be formed using collapsible cores, but they are not currently considered economically practical and should be avoided in the design.

Other functional and decorative features

Features such as logos, knurls, part numbers, and cavity identification marks can be easily molded in place. Sequential numbers are possible if tooling is justified and there is a suitable location on the component. Since these features must be ejected from the mold, they can not be located on surfaces that are parallel, or nearly parallel, to the direction of movement of the mold member on which they are located.

2.3.2.4 *Dimensional Precision and Surface Finish*

One of the strengths of the MIM process is its ability to control shrinkage of the green body during sintering to achieve close tolerances. In general, dimensional tolerances of ±0.3-0.5% of nominal dimensions can be held, with a lower limit of ±0.025 mm (0.001 in.). Typical angular tolerances are ±0.5°. Closer tolerances may be held on some features, but it may be necessary to relax tolerances on other features.

The very fine powder size and low porosity produce typical surface finishes of 0.8 micrometers (32 RMS).

2.3.3 P/F Shapes and Forms

The following discussion assumes that the reader is familiar with the basic principles of P/M compaction and the P/F process. If further information is required, please review sections 4.1.1 Compaction and 4.3 Powder Forging.

Figure 2.3-35 shows components that are being successfully made by P/F from one- or two-level preforms. The preforms are produced by P/M compaction and the design criteria discussed in that section apply. The number of levels in the preform must be consistent with the need to control density gradients in the preform. The P/F process can produce multiple levels from preforms with fewer levels because of the metal movement that occurs during the forming process.

Figure 2.3-35 P/F parts.

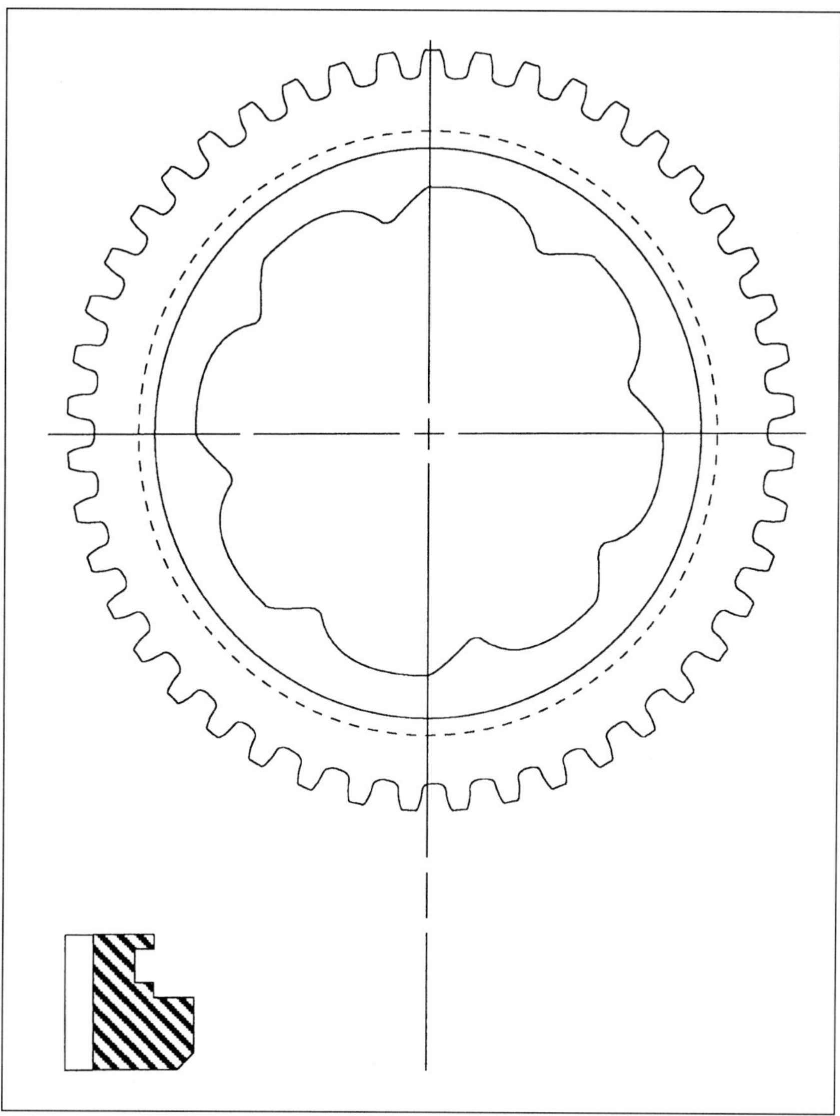

Figure 2.3-36 Typical ring gear for automatic transmission; fully finished.

Density gradients in the preform affect the response of the preform to the forging operation. During forging, areas of the preform with higher density (lower porosity) exert more force on the forging tools, which tends to displace or even cause local deflections in the tool, generating dimensional variations in the forging.

The effects of the forging process on the preform and the features of the component can be illustrated by following the steps in the design development of a typical P/F component, such as a ring gear for an automatic transmission. The example is hypothetical and is used to illustrate one of several ways that components of this shape are being made. Figure 2.3-36 shows the ring gear as it might be originally drawn by the design engineer. The ring gear has shallow teeth on the outside diameter, and a series of ramps for a one-way roller clutch on the inside diameter.

The P/F parts producer designs the two-level preform shown in cross section in Figure 2.3-37, and compacts and sinters it to a uniform density that is typically between 80% and 85% of the pore free density, depending on anticipated forging parameters.

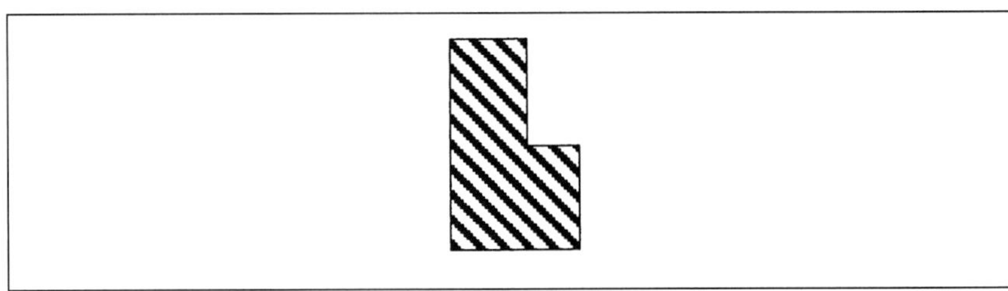

Figure 2.3-37 Cross section of preform for transmission ring gear.

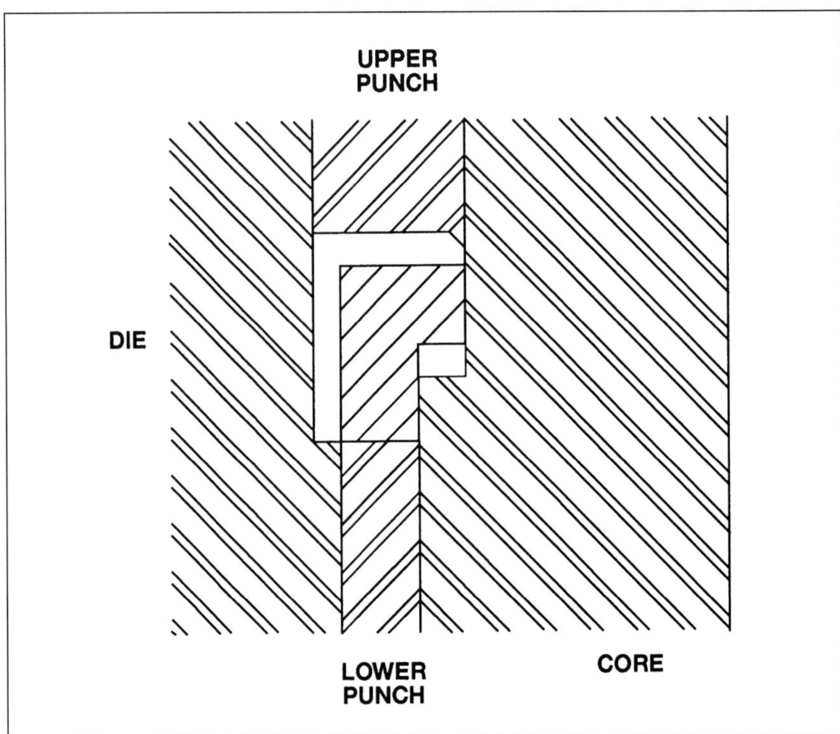

Figure 2.3-38 Transmission ring gear in the forging die prior to forging.

The preform is forged in a closed die operation. In this case, the teeth on the outside diameter are relatively shallow, and the preform does not require a large amount of metal movement to fill the cavity. The P/F producer may choose either to form the teeth in the preform or form them in the forging tool. In this example, they are formed in the forging tool to illustrate the effects of metal flow in the tools. The ramps on the inside diameter are deeper and require a significant amount of metal flow. Features of this size are made to nearly final form in the preform, as shown in the illustration.

Figure 2.3-38 shows a cross section of the ring in the tool. The ring is given its final shape in the cavity formed by the upper punch, die, lower punch and core. In the forging sequence, energy is imparted by downward movement of the upper punch, which subsequently retracts. The lower punch then pushes the ring out of the tool.

The forging operation reduces the height (dimension in the direction of forging) of the preform and forces metal into the recesses on the outside diameter, forming the teeth. The chamfer on the inside diameter is formed by a protrusion on the upper punch. The forging operation also brings all features, including the preformed ramps, to their final toler-

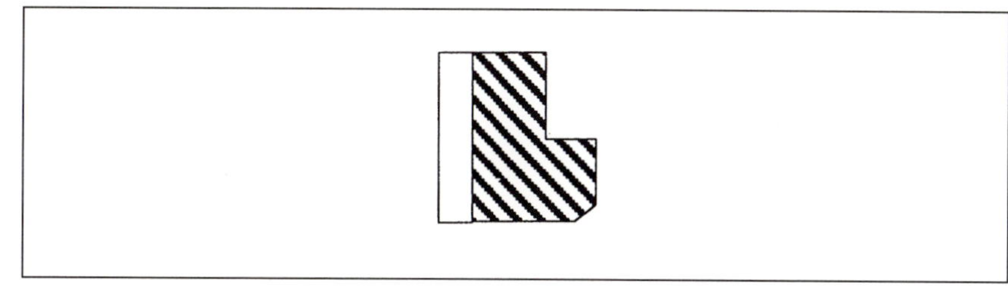

Figure 2.3-39 Cross section of transmission ring gear after forming and before finish machining.

ances and densities. The undercut cannot be forged and is machined. Figure 2.3-39 shows a cross section of the ring as forged.

The figure also illustrates features in this type of component that facilitate forging.

- A low angle (departure from the forging plane) is preferred on the I.D. chamfer, which is formed by a protrusion on the upper punch. As the angle is increased, the protrusion becomes larger and is more subject to breakage.
- Radii on inside corners of the forging, as large as feasible, promote flow of metal around corners in the tool and promote complete fill of all details. Radii also reduce wear and breakage of features in the tool.
- Radii of at least 1 mm should be provided on outside corners of the forging because sharp corners on the corresponding inside corners of the tool are difficult to fill. If a sharp corner is required, 1 mm of extra stock should be provided and the surface ground to eliminate the radius and sharpen the corner. A radius on the outside corner on the forging also reduces stress concentration in the tool.

Several guidelines that apply to conventional precision forging also apply to P/F. For example:

- The shape of the forging should be such that, when placed in the dies, the lateral forces will be balanced. Otherwise, extra features must be added to the dies to resist unbalanced lateral forces, increasing both the initial cost and maintenance costs of the dies. Shapes that are symmetrical, such as connecting rods and shapes that are axisymetric (or nearly so), such as the one in the example, do not cause this type of problem.
- Zero draft is possible on surfaces that are formed by the die and core rod, but not by the upper punch. The upper punch must release from the forging when it is retracted, whereas the forging is pushed off of the die and core rod by action of the lower punch.
- Repeated forging cycles cause progressive wear on the dies. Since die maintenance is driven by wear and tolerance requirements, costs arising from die maintenance will be contained by making design tolerances as generous as possible. Wear is not consistent throughout the die, factors such as pressure and metal flow increase wear.
- Re-entrant angles (undercuts) cannot be forged. Where required, they must be formed by machining operations.

The tolerances that the designer can anticipate are driven by design details and process variables, and are application specific. For example:

- Axial tolerances—in the direction of forging—are driven by variations in the mass of metal in the preform. Lateral tolerances are driven by metal flow as the cavity fills.

- If the preform has areas of higher density on one side than on the other, the preform will exert more lateral force on that side of the punch and deflect it toward the side with lower density. This deflection will make the inside diameter slightly eccentric with respect to the outside diameter.
- Eccentricity of the forging is driven by both the eccentricity and density of the preform. The eccentricity of the forging will be approximately twice that of the preform.
- Final dimensions of the forging are affected by thermal shrinkage as it cools to room temperature from a forging temperature typically around 980°C (1800°F). As features of the forging become more complex, shrinkage becomes more difficult to predict.
- Tool wear is highest in areas where there is the greatest amount of plastic flow in the preform.

Table 2.3-3 shows tolerances on P/F components that have been commercially produced. Actual tolerances should be determined on a case-by-case basis in consultation with a P/F parts producer.

Table 2.3-3 Typical Tolerances on P/F Components[a]

	Millimeters		Inches	
Parameter	Dim.	Tol.	Dim.	Tol.
Inside diameter	38.1	0.20	1.50	0.008
Inside diameter	63.5	0.25	2.50	0.010
Outside diameter	50.8	0.13	2.00	0.005
Outside diameter	50.8	0.25	2.00	0.010
Outside diameter	50.8-76.2	0.13	2.00-3.00	0.005
Outside diameter	25.4-50.8	0.10	1.00-2.00	0.004
Outside diameter	76.2	0.38	3.00	0.015
Outside diameter	95.2	0.25	3.75	0.010
Outside diameter	203	0.51	8.00	0.020
Concentricity	95.2	0.10	3.75	0.004
Roundness	95.2	0.10	3.75	0.004
Spline	25.4	0.23	1.00	0.009
Thickness	15.8	0.25	0.62	0.010
Thickness	25.4	0.38	1.00	0.015
Thickness	25.4	0.25-0.63	1.00	0.010-0.025

[a]Data are taken from Pease, L. Assessment of Powder Metallurgy Today, and Its Future Potential, Paper No. 830142, Passenger Car Meeting, Society of Automotive Engineers, Warrendale, PA., 1983.

2.4 DESIGNING P/M, MIM AND P/F COMPONENTS TO MEET STRUCTURAL CRITERIA

The MIM and P/F processes produce nearly fully dense components, which respond to applied loads in a manner similar to their cast and wrought counterparts. However, conventional P/M components contain microporosity, and respond differently to loads and test procedures. Therefore, conventional techniques for structural analysis must be modified somewhat. This section focuses on structural criteria for P/M components.

Conventional formulas for stress and strain are commonly employed to optimize the design of components such as gears and splines. Analytical methods can often predict product performance accurately, but may be impractical where inputs are not quantified or, when the component is complex and a cycle of testing and modification is relatively simple.

Mechanical properties of P/M alloys are compiled from tests performed on specially prepared specimens. The properties of the specimens may vary from actual components primarily because the test specimens are manufactured to minimize or eliminate density gradients, which may be significant factors in the components. Component design parameters that cause density gradients may cause the mechanical properties in some features to vary somewhat from tabulated data and from other features in the same component. The most critical parameter is the ratio of length (dimension parallel to the direction of pressing) to cross section, usually designated l/d. (See Section 4.1 for a further discussion.)

2.4.1 Tabulated Mechanical Properties

The structural performance of a P/M component usually depends on several mechanical properties either alone or in combination such as tensile and compressive strengths, modulus of elasticity, fatigue strength and impact energy.

1. **Tensile Strength.** Tabulated data for tensile strength displays yield strength based on 0.2% offset for ferrous alloys that are not heat treated and for nonferrous alloys. Ultimate tensile strength is displayed for heat treated alloys because they exhibit little difference between ultimate tensile and yield strengths.

    Tensile strength is directly affected by material density. Features with low l/d may be expected to achieve design strengths indicated by published stress data. Features with high l/d should not be designed to published stress levels unless the design anticipates that critical features will be revised as needed based on test results. When components are designed with low l/d in critical features, such as gear teeth, the designer can use conventional design formulas. Where strength of a feature is critical, it is essential to work with the P/M manufacturer to optimize the design for manufacturing, determine the strength that can be reasonably specified, and establish the test and inspection procedures for maintaining performance.

2. **Modulus of Elasticity.** Modulus of elasticity is directly affected by material density. This factor must be considered when applying finite element analysis (FEA) to the design of P/M components, particularly those with significant variations in density. Conventional methods for computing stress also assume constant elasticity. This factor should be borne in mind when results of stress calculations are interpreted. Poisson's ratio has been tabulated, and can be used in FEA.

    Tabulated values for modulus of elasticity indicate lower values for P/M alloys than for equivalent wrought and cast alloys, higher values than for die cast alloys, and much higher values than plastics and most composites. Higher elasticity may allow opportunities to reduce wall thicknesses and eliminate reinforcing features, such as gussets and ribs, when redesigning plastics, composites and die castings for P/M.

3. **Fatigue.** Ferrous P/M alloys exhibit an endurance limit, as do cast and wrought alloys of similar composition. The endurance limit increases with increasing component density. Fatigue performance of a P/M component relative to tensile strength is reasonably consistent. Therefore fatigue performance is usually verified by testing production samples, then developing a corresponding static test for the critical area as a convenient, reliable test.

4. **Impact Energy.** Impact requirements are very difficult to predict by analytical methods because the inputs are usually difficult to quantify. Published values of impact energy are useful for predicting the relative performance of two alloys in a given application. The impact energy of ferrous P/M alloys can be improved by increasing density and by infiltrating with copper.

5. **Bearing "K" Factor.** The "K" factor, which is important when designing bearings to withstand the forces associated with interference fits, is a measure of the theoretical radial crushing force that the bearing can withstand. The factor is developed by placing a diametrical test specimen between two flat surfaces and applying a load normal to the direction of the longitudinal axis (the radial direction). The point at which the load drops due to the first crack determines the K factor.

   The design radial crushing force is calculated from the "K" factor by the formula:

   $$P = \frac{KLT^2}{D-T}$$

   Where: P = radial crushing force, lb.
   K = "K" factor, lb/in.^{2}
   L = bearing length, in.
   T = wall thickness of bearing, in.
   D = outside diameter of the bearing, in.

Note: this equation becomes invalid when the wall thickness exceeds 30% of the outside diameter (T > 0.30D)

2.4.2 Other Structural Properties

Two other structural properties relevant to P/M components, which are not readily handled by standard methods, are creep and dent resistance.

1. **Creep.** Creep is very slow plastic deformation occurring at stress levels below the yield point. The temperature at which creep begins is generally indicated by the melting point of the alloy; the higher the melting point, the higher the threshold temperature. The operating temperature ranges for most P/M copper, iron, steel and stainless steel components are rarely above the creep threshold. If creep range temperatures are anticipated, alloy formulation can be modified to improve creep performance.

   P/M components may offer economical alternatives to injection molded plastics, composites and die castings designed for temperatures to 205° C (400° F). The most critical areas are points of concentrated loads. For example, a redesign for P/M may require fewer or smaller fasteners, require less thread depth in tapped holes, allow threaded fasteners in tapped holes rather than through bolts and nuts, and eliminate the need for steel inserts to distribute concentrated loads.

2. **Dent Resistance.** Dent resistance is the ability of a component to withstand impact without permanent deformation. A component subjected to impact stores energy by deflecting under load; when stress levels do not exceed the elastic range of the material, no permanent deformation occurs. The energy storage capacity within the elastic range of the alloy is proportional to the stress and the strain at the elastic limit. Dent resistance is distinct from impact resistance, which is the total energy of deformation through the plastic range to the point of rupture.

   Since stress is proportional to strain in the elastic range of P/M alloys, the energy absorbed within the elastic range is indicated by the yield strength and elasticity. For a test specimen under uniform tensile stress, the elastic energy absorbed is proportional to:

$$Y_S^2/(2E)$$

Where: Y_S^2 = yield stress
E = modulus of elasticity

The formula indicates that dent resistance can be enhanced by selecting an alloy with high yield strength and low elasticity.

2.4.3 Temperature Effects

The mechanical properties of all metals are affected at extremes of temperature from ambient. The temperature ranges in which most P/M components operate have little effect on copper, iron, steel and stainless steel alloys. In those applications where anticipated temperatures are high or low enough, alloy formulation can be modified to improve the properties.

The stability of mechanical properties at elevated temperatures often makes P/M an economical alternative to injection molded plastics, composites and die castings, which can experience loss of strength and reduced elasticity at approximately the same temperatures that induce creep. When redesigning for P/M, design economies discussed previously for elasticity, strength and creep may be applicable.

2.5 P/M BEARING APPLICATIONS

MIM and P/F offer bearing properties that are equivalent to their cast and wrought counterparts. However, the microporous structure of P/M components gives them distinct advantages for bearings or for applications where bearings can be integrated into the component.

Self-lubricating bearings are one of the oldest P/M applications (Figure 2.5-1). Continued progress in materials and processes has maintained the competitive position of P/M. When bearings are integrated into the component, the number of parts is reduced, cost is reduced, product function is improved and recycling is facilitated because the bearings do not have

Figure 2.5-1 P/M bearings.

to be removed for separate processing.

There are two distinct operating modes for bearings, depending on lubrication conditions: full film and partial film. Most full film bearings are hydrodynamic, which are characterized by an excess supply of lubricating oil, close shaft-to-bearing clearances, high speeds and high loads. Microporosity, which enhances P/M performance under partial film conditions, interferes with the formation of the continuous film required in hydrodynamic bearing applications.

Bearings operating at partial film (boundary layer) lubrication experience some metal-to-metal contact between the shaft and bearing during operation, making it essential to supply the bearing with lubricant. The ability to add liquid lubricants such as oil and solid lubricants such as graphite gives P/M an advantage over other processes when partial film bearings are required. The oil-holding capacity of P/M bearings usually provides sufficient lubrication for the life of the product. Where operating conditions require additional oil supply, reservoirs, wicks or oil-soaked felt washers are used.

Oil is introduced by impregnation after compacting and sintering. Impregnation is achieved by vacuum techniques or by soaking the components in heated oil; the amount of oil absorbed depends on the percent of interconnected microporosity. When friction generates heat, the oil expands and flows to the bearing surface, providing a self-lubricating effect. Upon cooling, it returns into the pores by capillary action.

Graphite to be used as a solid lubricant is mixed with the powder prior to compacting. The sintering process for iron components is controlled to inhibit the graphite from combining with the iron.

2.5.1 Bearing Alloys

Five properties are normally considered in selecting a material for a bearing:

1. Chemical compatibility with the lubricant
2. Thermal conductivity to dissipate heat
3. Mechanical properties to maintain structural integrity and withstand press-fit installation forces

4. Anti-galling with the shaft material
5. Wear characteristics with the shaft material

Most available lubricants are compatible with the P/M alloys used in bearing applications; thermal conductivity of the alloys is generally adequate. The other properties are achieved by selecting the appropriate P/M alloy from among six alloy groups. The strength criterion for a P/M bearing is the "K" factor, which is defined in Section 2.4.1, rather than ultimate tensile or yield strength.

The following discussion gives a brief overview of the alloy groups and application types. Detailed information on alloy properties is given in MPIF Standard 35 Material Standards for P/M Self-Lubricating Bearings.

1. **Bronze**

    Bronze offers excellent corrosion resistance, ductility sufficient to permit staking, and resistance to shock loading. Strength is generally adequate for use in components with integrated bearings. Typical applications are fractional horsepower motors, business machines, farm implements, hardware and machine tools. Graphite is often added to enhance lubrication. Bronze bearings with the addition of graphite are selected for conditions of heavy loads, oscillating and intermittent rotation and high temperature applications. They are generally selected where operating conditions impair the capability to develop lubricating film strength in the bearing.

2. **Iron and Iron-Carbon**

    Plain iron at densities of 5.6 to 6.0 g/cm^3 (71 to 76% dense) can be used as bearing materials for medium loads. The alloy is typically harder and stronger than 90-10 bronze. Combining carbon with iron produces a steel bearing that is stronger than pure iron, with higher radial crushing strength, greater wear resistance and higher compressive strength. Bearings with a combined carbon content greater than 0.3% may be heat treated to further improve mechanical properties.

3. **Iron-Copper**

    Copper is mixed with iron to improve strength and hardness; copper additions of 2%, 10% and 20% are common. At 20% copper the alloy is harder and stronger than 90-10 bronze and exhibits good shock loading capability. Iron-copper alloys are often used in applications requiring combined structural and bearing characteristics.

4. **Iron-Copper-Carbon**

    Carbon additions in amounts of 0.3 to 0.9% greatly strengthen iron-copper alloys. Carbon also allows further improvement of mechanical properties by heat treatment. This alloy group offers high wear resistance and high compressive strength.

5. **Diluted Bronze**

    Bronze is diluted with 40 to 60% iron to lower raw material cost. The bearings usually contain 0.5 to 1.3% graphite for self-lubrication. They are used as an economic alternative to bronze in light to medium load and medium to high speed applications such as fractional horsepower motors and appliances.

6. **Iron-Graphite**
   Graphite is mixed with iron and sintered to a low combined carbon content so that most of the graphite is available as lubricant. The bearings may be impregnated with oil to improve self-lubrication. The bearings also exhibit excellent damping characteristics and quiet operation.

2.5.2 Design of P/M Bearings

The design of P/M bearings and the selection of the optimum alloy depend primarily on shaft velocity and bearing load. Additional factors include type of operation such as stop-and-start, shaft surface finish, type of lubricant, and conditions of uneven loading. MPIF Standard 35 P/M Bearing Material Properties gives a systematic guide in the section Engineering Information for P/M Self-Lubricating Bearings. The discussion includes the following topics:

1. Recommended loads as a function of shaft velocity for five alloys
2. Environmental factors that can reduce permissible loads
3. Recommended press fits
4. Running clearances
5. Dimensional tolerances for plain cylindrical bearings

The engineering information in Standard 35 has proven helpful in designing bearings. Values are generally valid, but specific applications can cause exceptions. The design should ultimately be performed in consultation with a PMPA member specializing in the manufacture of P/M bearings.

2.5.3 Designing Bearings into P/M Components

Many of the P/M alloys in the six groups mentioned in Section 2.5.1 offer a combination of mechanical and bearing properties that allows the designer to incorporate bearing surfaces into the component, eliminating the need for separately fabricated and installed bearings. Integrating bearings imposes the following design restrictions:

1. Bearing areas should have a minimum of 10% interconnected microporosity to provide sufficient oil capacity.
2. Infiltration is not permissible in the bearing areas because the infiltrant fills the pores.
3. Machining operations if required on bearing surfaces should be performed so as to avoid sealing surface microporosity.

It is essential for the designer to consult with a PMPA member early in the design process. For example, by integrating design with tooling requirements it is often possible to vary density in the component, maximizing density in areas requiring strength and reducing density in bearing areas to increase oil holding capacity.

2.6 SURFACE TREATMENT

Practically all common methods of surface treatment that are used for wrought and cast products can be applied to P/M, MIM and P/F components. Some of the processes are described in Section 4. The following discussion is a brief guide for specifying the optimum surface treatment system based on product objectives. Since MIM and P/F components are free from interconnected microporosity, no process modifications are required. The microporous structure of P/M requires modifications, as noted in the following discussion.

2.6.1 Electroplating

P/M components can be plated with the same metals as cast and wrought components (Figure 2.6-1). For example, copper, nickel, chromium, zinc and cadmium plating are commonly specified. Electroless nickel plating also can be used, and peen (mechanical) plating is applicable to nonimpregnated components. P/M components should have microporosity sealed to avoid entrapment of plating solutions in the pores. Burnishing, buffing, rolling, peening, steam treating, impregnation and infiltration are employed to seal surface pores.

Impregnation and infiltration completely close surface microporosity. However, the cost of infiltration is normally prohibitive unless it is done for other purposes such as improved pressure tightness, strength or fracture toughness. The most commonly used impregnants currently specified are polyester and anaerobic types. Mechanical methods—burnishing, buffing, rolling and peening—and infiltration are sometimes unacceptable because they affect dimensional tolerances.

1. Copper Plating

   Acid copper plating generally gives the best result on P/M components because of its excellent throwing power. When the copper color is desired, a clear coating such as lacquer applied over the bright copper is specified to prevent the copper from oxidizing. Most copper plating is performed as a base for either nickel or nickel and chromium.

2. Nickel Plating

   Nickel, applied either alone or over a copper base, is used to enhance wear resistance and to provide a bright, corrosion resistant surface, as a final coat or as a base for chromium. It is applied to steel and brass.

Figure 2.6-1 Examples of components surface treated by blackening and zinc dichromating.

3. Chromium Plating

Either decorative or hard chromium plating may be applied to P/M components. Decorative chromium is usually deposited to a thickness of 0.00025 to 0.00050 mm (0.00001 to 0.00002 in.). Hard coating is a heavier deposit, which is applied for wear and heat resistance.

In many applications, P/M components can be oil dipped after plating to eliminate the need for the corrosion resistant underlying nickel base that would be required for cast or wrought components. The oil film covers the entire plated surface and provides excellent corrosion protection for the chromium plate.

Multiple plating is often specified to achieve wear resistance. Copper in thicknesses up to 0.13 mm (0.005 in.) is first used as a leveler and sealer. A nickel layer ranging in thickness from 0.13 to 0.25 mm (0.005 to 0.010 in.) provides the support, hardness and corrosion resistance for the chromium. The chromium thickness is typically 0.00025 to 0.00075 mm (0.00001 to 0.00003 in.).

4. Zinc and Cadmium Plating

Zinc and cadmium plating are used for corrosion resistance. Cadmium plates faster and has greater throwing power than zinc. However, the cost of cadmium is typically ten or more times that of zinc. Zinc will withstand greater concentrated salt spray, but the two perform equally in exposures 80 feet from the ocean. Cadmium has a lower coefficient of friction with steel than does zinc.

Cadmium is used indoors on radio and electronic instruments because the products of corrosion are lower in volume than those associated with zinc. It also bonds better with solder. Cadmium is never used with food processing equipment because of possible toxic effects.

2.6.2 Resin Impregnation

The designed microporosity of P/M components permits their impregnation with resins. Solid resin impregnants are used for the following purposes:

1. Provide lubrication
2. Develop pressure tightness
3. Permit the use of processing solutions (such as plating solutions discussed above) that would otherwise be trapped in the pores and degrade the component
4. Improve machinability

Resin impregnation in some cases significantly improves machinability; however, impregnants cannot withstand heat treating temperatures.

2.6.3 Mechanical Methods of Surface Treatment

Mechanical methods of surface treatment, including liquid blasting, glass beading, wire brushing, belt sanding, tumbling, and vibratory finishing, are used to:

1. Clean components
2. Impart a smooth, bright finish
3. Eliminate burrs
4. Remove sharp corners

These processes can be adapted to minimize the closing of surface pores, or they can be used to close surface pores. However, processes that close surface pores can affect dimensional tolerances. When a process requires the use of abrasives, there is potential for abrasive particles to lodge in surface pores. The particles can cause excessive wear on cutting tools and they can be detrimental to the functioning of the components.

Shot peening is used to improve the surface fatigue properties of P/M components, such as gear teeth. It also accomplishes the four objectives enumerated above.

2.6.4 Processes Specified for Ferrous Components

Blackening or bluing, and steam treating are specified only for ferrous components. Furnace blackening is specified to give indoor corrosion resistance. Oil dipping improves corrosion resistance and produces a deeper color. An oil that leaves a dry film on the components may be specified.

Chemical blackening using a commercial liquid salt bath can also be specified. Components should be impregnated with a resin that will not break down in the bath. An oil dip can be specified to improve appearance and corrosion resistance as with furnace blackening. Nickel and copper in ferrous alloys tend to adversely affect most blackening baths and seriously affect components color.

Steam treating is commonly specified to increase wear and corrosion resistance. It additionally increases component density and compressive strength. The layer of black oxide, (Fe_3O_4) which forms on all particle surfaces including those exposed to interconnected microporosity, produces a surface that is smooth, and sometimes reflective. Oil dipping after steam treating can be specified. There are, however, four potential disadvantages of steam treating.

1. There is a slight size change, up to 0.0075 to 0.0125 mm (0.0003 to 0.0005 in.) increase per surface.
2. Components are somewhat less ductile and more difficult to machine.
3. Some steam treating cycles can cause significant reductions in tensile strength, elongation and impact resistance with iron-carbon and iron-copper-carbon steels.
4. When heat treated components are steam treated, steam treatment temperatures of 480 to 600°C (900 to 1100°F) negate some of the effects of heat treatment.

2.6.5 Processes Specified for Stainless Steel Components

Thermal passivation is used to improve the corrosion resistance of stainless steel components. The process, which consists of a furnace heating cycle in air at 325 to 500°C (620 to 930°F) for thirty minutes, imposes no design or process limitations on the component. The cycle generates a film, which passivates the component surface into the pores, increasing resistance to certain corrosive agents and improving resistance to pitting corrosion.

2.6.6 Processes Specified for Aluminum Components

Aluminum is considered difficult to electroplate because of its affinity for oxygen. Chemical conversion coatings of chromates, phosphates and aluminum oxide are specified to improve corrosion resistance and impart gold, green or gray colors.

Anodizing is specified for corrosion resistance using either chromic or sulfuric acid processes. Hard coat anodizing is specified for both corrosion and wear resistance using sulfuric-oxalic acid. Coloring agents, either organic or inorganic, can be applied with either process. Surface sealing may be required to prevent entrapment of processing solutions, depending on component density.

2.6.7 Other Surface Treatment Processes

Other specialized and proprietary surface treatment processes are used with P/M components to gain specific product advantages.

2.7 PROTOTYPING

Before a product is released to the marketplace, performance must be verified. Since material properties are affected by manufacturing processes, final verification must be performed on products made from production tools operating under production process controls. When a product is being redesigned from an existing functional design, the performance of the existing design may enable the designer to redesign with a high level of confidence. In those cases, analytical techniques such as FEA are employed, a set of production tools is fabricated, and prototypes are produced with no intermediate prototyping stage. Analytical techniques are impractical when inputs can not be quantified, or the component is very complex and a cycle of testing and modification is relatively simple. In some cases, the optimum procedure includes both analytical techniques and prototyping.

Prototypes are fabricated and tested to evaluate one or more of the following characteristics:

1. Clearances, installation problems, or assembly problems
2. Appearance
3. Environmental and galvanic corrosion
4. Strength and rigidity
5. Fatigue life
6. Impact energy
7. Bearing properties
8. Wear resistance
9. Filtering characteristics (P/M only)
10. Manufacturability

2.7.1 Prototyping Options for P/M.

Three prototyping options are commonly employed for P/M.

Option 1-machine from a P/M blank. P/M blanks, as pressed and sintered or infiltrated, are available from P/M component manufacturers for fabricating prototypes. Machining has a minor effect on surface microporosity. The blanks closely approximate the mechanical and physical properties of the production component except that density is relatively consistent throughout a simple blank, but more pronounced gradients may occur in a production component either by design or, unavoidably, due to complexity. This option is inexpensive and fast, but is limited by the sizes of available blanks.

Option 2-press from a set of P/M tools that contain defined features, sinter, and then machine the remaining features. For example, a component that includes both a spline that is defined and gear teeth that must be defined by test can be made from tools that form the spline, plus a blank from which the gear teeth are cut.

Option 3-press the entire component from a set of P/M tools and sinter. This option is viable when the component configuration is relatively simple or when a large number of prototypes is required. It is normally the most expensive and time consuming option, but the most reliable because component properties are similar to those of the end product.

2.7.2 Selecting the Proper Prototyping Option for P/M

The option selected depends on the purposes for prototyping.

1. Clearance studies, installation problems, and assembly problems require a prototype with the correct configuration and provisions for attachment (tapped holes, set screws, etc.). All options are generally satisfactory.

2. Appearance studies usually imply painted or plated surfaces; therefore surface quality is important.

3. Environmental and galvanic corrosion depend primarily on the chemical composition of the alloy, porosity, and surface treatment. Options 1 through 3 are generally satisfactory. However finish machining operations must be performed so as not to affect surface microporosity in a way that will influence the surface treatment.

4. The strength of a P/M component is determined by factors such as material composition, density, infiltration (where applicable), and heat treatment. Rigidity is determined by material composition, infiltration (where applicable), and density. All options can be employed to produce useful prototypes for initial strength and deflection studies. However, prototype features that are machined from sections pressed to a simple configuration, such as a P/M blank, may have less pronounced density gradients than in production components.

5. Fatigue performance is determined by the material composition, surface finish and the internal structure of the component, particularly the microstructure. Prototyping criteria are similar to those for strength.

6. Impact energy is not readily evaluated by analytical techniques such as FEA in many applications because the inputs are difficult to quantify. Therefore tests are often required. Prototyping criteria are similar to those for strength.

7. Bearing properties depend on material composition and microporosity. Where bearing properties are the only properties to be evaluated, all options are generally satisfactory. Where strength, fatigue or impact energy are also to be evaluated, prototyping criteria are similar to those for strength.

8. Filtering characteristics, similar to bearing properties, depend on material composition and microporosity. Since the size of the interconnected pores is important, as well as the percent microporosity, the designer should specify the maximum size particle that can pass through the filter, and work with the P/M manufacturer to achieve the minimum pressure drop. Option 1 is the pro-

totyping option most often selected. Option 2 is also applicable; however, special machining techniques are required to avoid displacing surface metal and closing the pores.

When anticipating a prototyping program, it is advisable to consult a PMPA producer for advice in matching properties of the prototype to those of the production component.

2.7.3 Prototyping Options for MIM and P/F

The very low levels of microporosity in MIM and P/F components develop properties that are very close to that of cast or forged components made from alloys with similar chemical composition. In most cases, product development proceeds from theoretical analysis to prototypes made from production tools. Prototype production runs are used to establish process variables.

Where the cost of prototype tools must be contained, several methods are available for approximating performance of MIM and P/F components at sharply reduced tooling cost and lead time. When only the first three characteristics listed above are of concern, any fabrication method that achieves the configuration in the desired alloy may suffice. Properties can be closely approximated at somewhat higher tooling cost and increased lead time by casting or forging to rough shape and finish machining critical features. MIM components can also be prototyed by machining from MIM blanks. In all cases, it is advisable to consult with the MIM or P/F parts producer for guidance in manufacturing and testing prototypes.

2.8 GUIDELINES FOR SPECIFYING P/M

The designer and the P/M parts maker are partners in profits. The partnership is enhanced when it begins in the design and development phases. A partnership formed early in the design stage with a PMPA member firm[3] is the best way to take advantage of the latest techniques in a rapidly growing, technologically advancing industry.

When the design reaches the point where requests for quotations are issued, accurate and adequate specifications are essential. Accurate specifications require a working knowledge of P/M manufacturing processes (see Section 4) and materials, (see Section 3). For more detailed information on materials, properties and specifications, refer to MPIF Standard 35.[4]

The P/M process allows the latitude to minimize cost if specific component requirements are not defined more rigidly than necessary. Widely differing costs may result from varying quality levels, changes in tolerance or design, or failure to specify minimum properties. There may be more than one MPIF standard material that can satisfy the purchaser's functional requirements. Therefore, when requesting a quotation, the following guidelines should be observed:

1. **Quantities.** Adjust order quantities, annual usage and future estimates to take advantage of high volume cost reductions.

---

3   For a listing of MPIF member firms, obtain the current P/M Buyers' Guide from the Metal Powder Industries Federation.

4   MPIF Standard 35 is available from the Metal Powder Industries Federation.

2. **Application Area.** Note whether the P/M part will replace one currently in production, or if it is a new application. Specify the application area such as automotive, food processing, office equipment, military, aerospace, medical, etc.

3. **Design Drawings.** Submit detailed drawings of the component including assembly drawings and, where possible, samples of existing components or prototypes. Identify materials that have performed satisfactorily for the application.

4. **Allowable Variations.** Note whether the design can be modified without affecting function; if so, designate where.

5. **Physical, Mechanical, Corrosion and Dimensional Requirements.** Specify the necessary physical, mechanical, corrosion and dimensional requirements. Two areas require special attention.
   - Possible density variations in a finished P/M part make it advisable to locate critical apparent hardness measurements on the drawings. The P/M supplier and purchaser should agree on the hardness, measuring procedure, and hardness scale for each component tested. The surface condition should also be specified and agreed upon because polishing or machining can close surface voids and affect hardness readings.
   - It is highly recommended that a proof test or destructive testing method be established to ensure that the P/M components meet the application requirements. Special fixtures or subassemblies for use by both parties may be required. Minimum values or ranges of values may be determined theoretically or by actual testing of the first samples produced from production tooling. Such tests should supplement material specifications on the drawings.

6. **Surface Treatment.** Specify required finish such as plating, chromating or oxide coating.

7. **Service Conditions.** Describe anticipated service conditions such as heat, moisture, impact, or corrosive atmosphere.

8. **Finish Machining Responsibility.** If finish machining operations are required, specify what they are and whether they will be performed by the P/M parts supplier.

9. **Process Control.** The P/M process lends itself to statistical process control. The high volume production capability, repeatability and attractive cost benefits of P/M make it imperative to establish process capability, which should be agreed upon with the P/M parts manufacturer. To achieve the necessary controls for critical dimensions or functional tests such as torque, bending, and impact, it is recommended that identical sets of procedures and inspection fixtures or gages be provided by the purchaser (one to the P/M parts supplier and one retained by the purchaser). This will ensure that both evaluate the components in the same manner.

10. **Gears, Splines and Sprockets.** Specific data required for these components include:
    - Number of teeth
    - Theoretical pitch diameter
    - Pressure angle

- Measurement over wires
- Tooth thickness
- Tooth root radius
- Backlash
- Helix angle
- AGMA quality class

11. **Bearings.** Specific data required for these components include:
- Load
- Speed
- Ambient temperature conditions
- Continuous or intermittent service
- Shaft material and finish
- Bearing material
- Diameter and length
- Tolerance requirements
- Lubricant requirements

12. **Filters.** Filter performance depends on the size of the interconnected pores as well as the total per cent microporosity. The designer should therefore specify the maximum size particle that can pass through the filter, then work with the P/M component manufacturer to achieve the minimum pressure drop.

13. **Other.** The purchaser and P/M manufacturer should also agree on:
- Finished surface condition; e.g. remove surface oil for use in automatic assembly equipment.
- Surface finish specification and method of measurement, in view of surface finish effects on the mating component.
- Gauging methods; e.g "Go, No-Go" or direct reading.
- Critical characteristics requiring SPC.

Composition and Properties of Alloys and Materials

Metal powders are precisely engineered materials that meet a wide range of performance requirements. They are available in numerous types and grades designed for P/M, MIM and P/F processes. When fabricated into components, mechanical properties compare favorably with materials used in other metal working processes. Additionally, P/M, MIM and P/F offer wide latitude to develop engineered materials with properties tailored to the application through customary metal alloying. P/M materials have microporous structures with mechanical properties somewhat lower than wrought or cast counterparts, depending on the amount of microporosity. The controlled microporous structure of P/M also allows infiltration with an alloy of lower melting point to produce a component with a composite structure that can not be produced with conventional cast, wrought, MIM or P/F processes.

The engineering properties of P/M, MIM and P/F components are affected by several factors, including:

1. *Material type*
2. *Powder fabrication process*
3. *Component manufacturing processes*

The type of material determines the properties that can ultimately be achieved in the end product. These effects are discussed in Sections 3.2.1, Powder Materials for Structural Applications and 3.2.2, Powder Materials for Bearing Applications.

The powder fabrication process affects the particle size, size distribution and shape. These effects are discussed in Section 3.1 Powder Characteristics.

The component manufacturing process affects factors such as density, bond strength between particles and, in the case of P/F grain flow. These effects are discussed in Section 4 Manufacturing Processes.

3.1 POWDER CHARACTERISTICS

The techniques employed to fabricate metal powders for P/M, MIM and P/F processing are grouped into three general categories:

Atomization
  Water
  Gas
  Centrifugal
Chemical
  Reduction of oxides
  Precipitation from a liquid
  Precipitation from a gas
  Thermal (vapor decomposition and condensation)
Electrolytic

The technique used to fabricate a powder is determined by economics, the raw material, and the desired particle shape, size and size distribution. Particle shape, size and size distribution strongly influence the characteristics of powders, particularly their behavior during die filling, compaction and sintering. The range of shapes includes highly spherical, highly irregular, flake, dendritic (needle-like) and sponge (porous) (Figure 3-1).

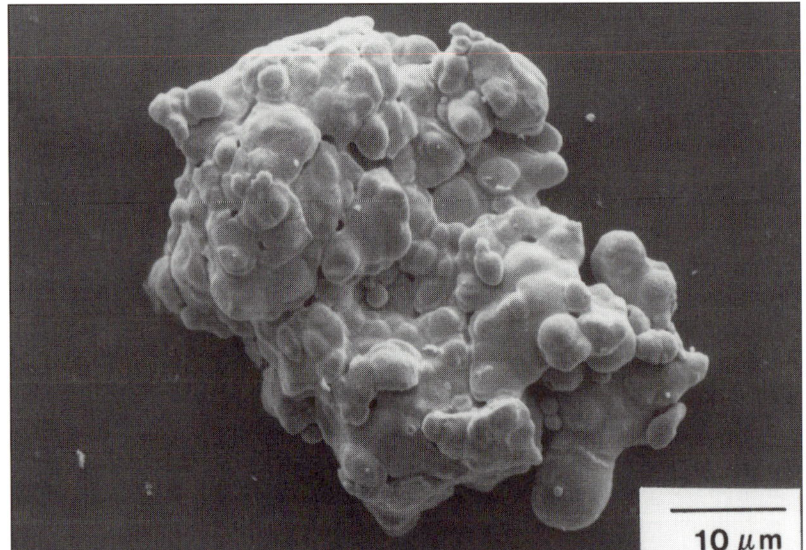

a.

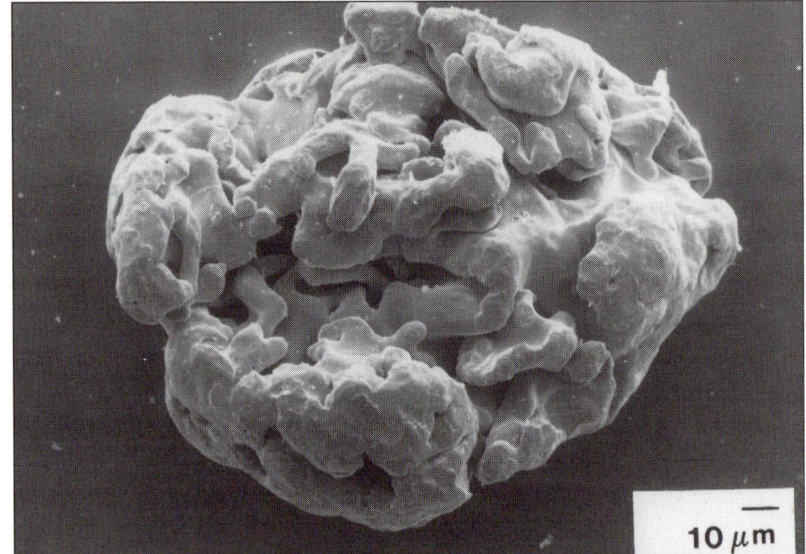

b.

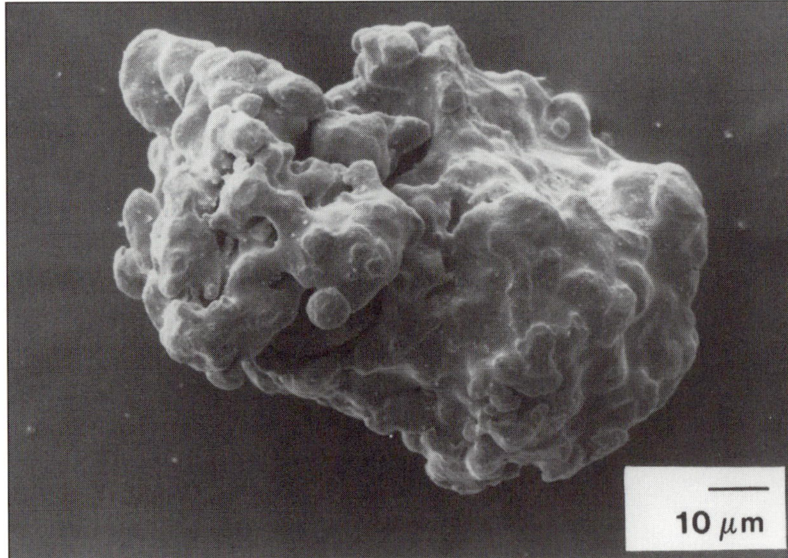

c.

Figure 3-1 Typical metal powder particle shapes: a. atomized copper; b. sponge iron; c. atomized iron.

It is desirable to have a broad distribution of particle sizes in the powder mix for P/M and P/F, to facilitate compaction. Smaller particles fill voids among larger ones, increasing packing density and improving material properties. However, fine particles tend to segregate from the mix unless appropriate procedures for handling are followed. Particle size is quantified in terms of size distribution or sieve analysis.[5] Four other quantifiable factors important to powder behavior are: apparent or bulk density, flow rate, compressibility and green strength.[6]

Metal powders are classified as elemental, partially alloyed or prealloyed. Elemental powders are used where high purity is essential to material properties, such as the magnetic properties of iron or the electrical and thermal conductivity of copper. They are also used to produce alloys by mixing with other powders prior to pressing and sintering. This procedure allows the P/M component manufacturer latitude to prescribe a mix that is tailored to design requirements. Elemental powders exhibit lower yield strength than prealloyed, allowing them to deform more easily during compaction, increasing density and green strength in the compact. However, elemental powders do not attain the degree of homogeneity of prealloyed powders during normal sintering. This condition generally results in lower fatigue performance and hardenability compared with prealloyed powders in forged components.

Partially alloyed powder is a metallic powder composed of two or more elements, which are partially alloyed in the powder manufacturing process. The alloying additives are diffusion bonded to maintain the compressibility of the base powder. These powders are non-homogeneous, and therefore have a microstructure that is quite diverse, containing a number of different, but controllable phases. The chemical composition can be controlled closely, which results in precise dimensional change properties. The materials have demonstrated excellent fracture toughness.

Prealloyed powder is a metallic powder composed of two or more elements that are produced by atomization of a liquid metal stream. The prealloyed particles are of the same nominal composition throughout.

The sintering process accomplishes alloying without gross melting. Alloying can also occur without melting when the diffusion occurs in the solid state, such as nickel in iron. A minor component, such as copper in iron, can melt while the major element remains solid. Each powder particle, in effect, becomes a micro ingot of alloy with a homogeneous chemical composition, producing a uniform microstructure in the compact. Prealloying is used to achieve alloys that are difficult to achieve by mixing elemental powders, such as stainless steel.

In most cases the powder mix contains elemental or prealloyed powder plus additives that serve purposes other than alloying. For example:
- P/M and P/F powder mixes use lubricants for a variety of reasons, such as increased tool life, and to facilitate ejection of the compact from the die.
- P/M powders sometimes use organic materials that occupy spaces between powder particles, then vaporize at sintering temperatures to develop controlled microporosity in applications such as filters.

---

5 MPIF Standard 05
6 MPIF Standards 04, 48, 03, 15 and 45

• MIM powders are mixed with an organic binder that allows the feedstock to be molded in a process similar to plastic injection molding, then is systematically eliminated in subsequent processing.

The Metal Powder Industries Federation (MPIF) has established a three part alphanumeric code for P/M, which designates powder composition and properties of the finished component. The prefix denotes the basic element and the most important alloying elements. The numerals following the prefix letters indicate the composition of the major and minor alloying elements. The suffix indicates minimum strength. A full explanation of the system and comprehensive documentation of material properties are contained in MPIF Standard 35. Examples are given for each type of alloy discussed below.

3.2 POWDERS USED FOR P/M

The characteristics of powders used for P/M are driven by the manufacturing processes and the application. Factors such as particle shape, size, and size distribution determine, in part, factors such as the strength of the compact, the achievable density and the response to heat treatment. The powder characteristics are also tailored to the application. Those used for structural applications have distinctly different characteristics from those used for bearings or filters.

3.2.1 Powder Materials for Structural Applications

Practically all metals can be made into powder. This discussion addresses the four most widely used metal groups in P/M structural applications: ferrous, stainless steel, copper and aluminum.

3.2.1.1 Ferrous

Iron: Iron powder is the most widely used P/M material for structural components. It is sometimes used alone, but most frequently with small additions, singly or in combination, to improve mechanical properties. Carbon, copper, nickel and molybdenum are the most common alloying elements.

A typical designation code indicating an iron base is shown in Figure 3-2. The code begins with the letter F. Subsequent letters indicate one or more alloying elements. The first two numerals following the letter prefix indicate the nominal content of the major alloying element in per cent by weight. The third and fourth indicate the combined carbon content as follows:

Code Designation	Carbon Ranges
00	0.0% to 0.3%
05	0.3% to 0.6%
08	0.6% to 0.9%

Ferrous alloys with 0.3% or more combined carbon (05 and 08 designations) can be hardened and tempered to increase strength, hardness and

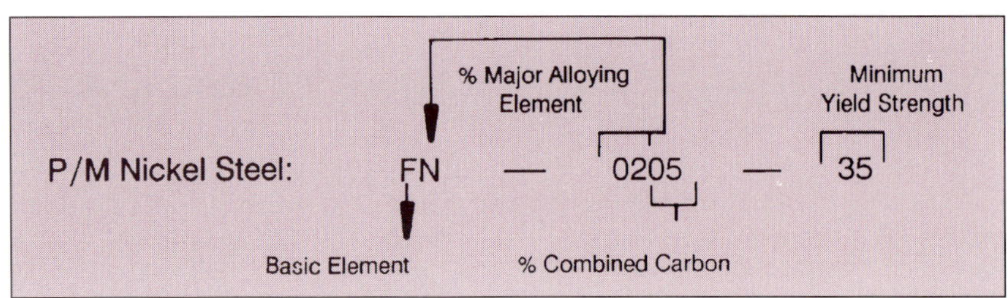

Figure 3-2 Typical MPIF designation code for an iron base material.

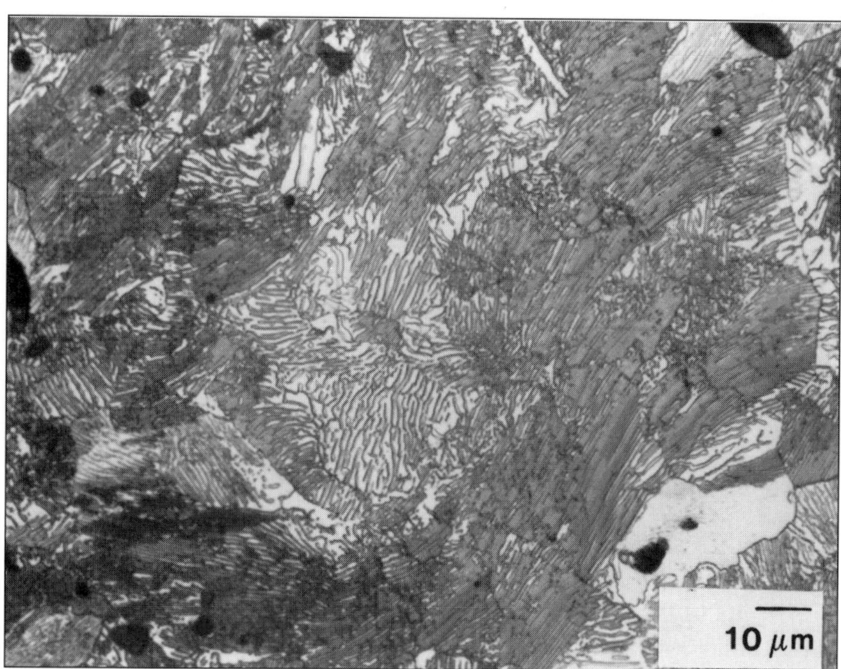

Figure 3-3 Photo micrograph of sintered steel P/M composition. Atomized iron, 0.8 C; density 7.1 g/cm³ (91% relative density).

wear resistance (Figure 3-3). Heat treated ferrous alloys have a four character suffix consisting of two digits indicating minimum ultimate tensile strength in thousand pounds per square inch, followed by the letters HT indicating heat treatment. Ferrous alloys that are not heat treated have a two digit suffix indicating minimum yield strength in thousand pounds per square inch.

It is important to note the distinction in minimum strength designations between ferrous alloys in the sintered and heat treated conditions. The yield strength of heat treated alloys is close to ultimate tensile; therefore minimum ultimate tensile is indicated. Ferrous alloys that are not heat treated exhibit yield strengths lower than ultimate tensile, and minimum yield strength is designated.

Carbon Steel: Plain carbon steels are made from mixtures of iron and graphite powders. Sintering diffuses the graphite into the iron to produce a steel structure. Low, medium and high density components produced from these powders range in tensile strength from 110 to nearly 415 MPa (16,000 to 60,000 psi). Strengths can be increased to over 620 MPa (90,000 psi) by heat treatment. Prealloyed low carbon steel powder is also available. A typical MPIF designation for an iron-carbon steel is F-0008-35. The F prefix designates iron, the first two digits indicate no major alloying element, the 08 indicates 0.6 to 0.9% combined carbon and the 35 indicates 240 MPa (35,000 psi) minimum yield strength.

Copper Steel: Addition of copper powder to iron powder increases strength and hardness. Copper steels contain from 1.5 to 10.5% copper and up to 0.9% carbon, depending on the application. Low, medium and high density components and bearings produced with these powders have tensile strengths from 140 to more than 550 MPa (20,000 to 80,000 psi). Heat treatment can develop strengths up to 690 MPa (100,000 psi). A typical MPIF designation for copper steel is FC-0208-80HT for a ferrous alloy with a nominal 2% copper, 0.6% to 0.9% combined carbon heat treated to 555 MPa (80,000 psi) minimum ultimate tensile strength.

Nickel Steel: Nickel powder additions of 2% to 4%, with or without copper powder, to iron powders develop tough, high strength components and excellent fatigue strength. Tensile strength for components not heat treated ranges between 345 and 515 MPa (50,000 and 75,000 psi). Heat treatment increases strength to as high as 1240 MPa (180,000 psi). A typical material designation is FN-0208-50 for a ferrous alloy with a nominal 2% nickel, 0.6% to 0.9% combined carbon and 345 MPa (50,000 psi) minimum yield strength.

Infiltrated Iron and Steel: Porous ferrous components can be infiltrated with lower melting materials, such as copper or copper alloys, to increase density and provide improved impact resistance. Copper content ranges from about 8 to 25%, depending upon the pressed density of the component. Residual microporosity is reduced to 4 to 8%. Standard copper infiltrated components have tensile strengths ranging from 450 to 620 MPa (65,000 to 90,000 psi). Heat treatment increases strengths up to 900 MPa (130,000 psi). Infiltration can additionally provide local density improvements where it may otherwise be difficult to achieve in the pressing operation. Eliminating interconnected microporosity permits plating without resin impregnation. A typical MPIF designation for infiltrated steel is FX-2008-60 for a copper infiltrated ferrous alloy with a nominal 20% copper, 0.6% to 0.9% combined carbon and a minimum yield strength of 415 MPa (60,000 psi).

Phosphorous Iron: Phosphorous is added to iron powder in small amounts to accelerate the sintering cycle, enhance electromagnetic properties and provide a unique combination of good ductility and strength.

Low Alloy Steels: Low alloy steels are made from prealloyed steel powders using nickel, molybdenum and manganese as the major alloying elements. Graphite powder is also admixed to provide the necessary level of carbon in the final material. Low alloy steels are typically used where high performance materials capable of being heat treated are required.

3.2.1.2 Electromagnetic Materials

DC and low frequency soft magnetic applications require materials offering good permeability, purity and density. P/M soft magnets show lower flux densities than wrought magnets of the same size and shape because of microporosity. The lower densities can be compensated through the use of pure, high compressibility phosphorous iron powders, and by designing the P/M components somewhat larger. Cost of the small additional amount of material is usually more than offset by P/M's elimination of scrap.

A variety of P/M materials are available for soft magnetic components.[7]

1. Low Density Iron, characterized by a wide hysteresis loop with a low maximum flux density (B_{max}) and high coercive force (H_c), is the least expensive.
2. High Density Iron, with characteristics approaching wrought iron, have a higher B_{max} and lower H_c, resulting in lower core losses.
3. Phosphorous Iron offers improved permeability (μm) and H_c. The phosphorous also minimizes the effects of oxygen, nitrogen and

---

7 See bibliography, Section 7

other contaminants.

4. Silicon Iron, with 3% silicon content, is used where high electrical resistivity results in lower core losses and fast magnetic response. The alloys are made by mixing a pure iron with an iron-silicon inter-metallic. These materials have B_{max} somewhat lower than iron, a low H_c, and considerably higher maximum permeability than iron. High temperature processing is necessary to achieve a homogeneous material, and the resulting improvements in magnetic properties.

5. Nickel Iron provides fast magnetic response for minimum applied field, and has the fastest permeability rise of any P/M material. Nickel iron components are generally made from 50/50 prealloyed powder and have very low H_c.

6. Ferritic stainless steels are utilized for magnetic applications that require only a moderate level of magnetic performance, but enhanced corrosion resistance.

Average expected values of magnetic properties, which are attainable in routine production, are shown in Table 3-4 in Section 3.5.4 Summary, at the end of this chapter. Actual values may be higher or lower, depending on process variables.

Another type of P/M magnetic component-high frequency cores—are produced by insulating iron powder particles with plastic coatings. They are pressed by conventional P/M methods and the plastic binder is "cured" at low temperatures. These iron powder cores are used in high frequency coils for numerous electronic applications and are closely allied to ferrite cores and the ceramic magnetic materials in electronic inductive components.

Permanent magnetic materials are fabricated by the P/M process to produce sintered Alnico, rare earth and bonded magnets.

3.2.1.3 Stainless Steels

P/M stainless steels are selected for applications where improved mechanical properties and a high level of corrosion resistance are required. Applications include automotive, appliance, hardware, marine, and food industry components and filters. Due to the presence of pores, sintered stainless steels have a large surface area compared with wrought stainless steels, and require careful process control. With such control, the corrosion resistance of sintered stainless steels can meet the 100 hour 5% NaCl salt spray test (ASTM B117), which is specified in many demanding applications. Sintering temperatures and atmospheres can be adjusted to affect impact energy, ductility and corrosion resistance.

The root of the MPIF designation is SS-, indicating stainless steel, followed by three digits indicating the grade. One or two additional characters are employed to designate minor distinctions within the grade. A suffix indicates minimum yield strength or minimum ultimate tensile strength and heat treatment, similar to the ferrous designation. Commonly used stainless steel P/M grades include:

1. SS-303 and 304 alloys are austenitic grades and are nonmagnetic. SS-303 is preferred for components requiring extensive secondary machining. SS-304 alloy is used as a general purpose grade.

2. SS-316 alloys are general purpose austenitic grades with the best combination of properties in a P/M stainless steel alloy. Strength and

hardness are high and corrosion resistance is superior to SS-303 and 304 alloys; machining is generally more difficult. The alloys are non-magnetic.

3. SS-410 is a standard martensitic grade used in applications requiring hardness and wear resistance. Carbon and nitrogen content can be adjusted for increased heat treatment response. Furnace cooling after sintering develops relatively high hardness, which can be increased with a secondary quench and tempering heat treatment. Corrosion resistance is fair and machinability poor. Without graphite addition, 410 alloys become ferritic with a simultaneous improvement in corrosion resistance. All grades in the 400 series develop maximum magnetic properties with decreasing levels of interstitials such as carbon and nitrogen.

4. SS-430 and 434 are ferritic grades. They are used in applications requiring magnetic properties. Corrosion resistance is superior to that of SS-410.

Because of their tendency for nitriding, which can occur upon sintering in dissociated ammonia, special suffixes have been included into the new stainless steel material designations to show the changes in mechanical properties with increased nitrogen alloying. The (N1) and (N2) suffixes indicate that the material was alloyed with nitrogen at levels exceeding 2000 parts per million. The (L) suffix denotes properties attained with very low nitrogen, oxygen and carbon contents. The (HT) suffix is applied to the 410 alloy, which is the highest strength stainless alloy, to represent the martensite grade. The 410 grade can also be supplied in the ferritic condition by following the sintering practice for the (L) grade stainless steels.

A typical stainless designation is SS-303N2-35, indicating a stainless steel grade 303 with N2 alloying providing a minimum yield strength of 240 MPa (35,000 psi) and 3.0% minimum elongation. This compares with SS-303L-12, which is the same chemistry, except that it is high temperature sintered in a nitrogen free atmosphere. Note that the yield strength was reduced and ductility increased to 12% elongation.

3.2.1.4 Copper Base

Copper base materials include pure copper, brasses, bronzes and nickel silver. The designation code is similar to the ferrous group with three exceptions.

1. The first character of the prefix is C, indicating copper.
2. The four digits following the prefix indicate the nominal percent content of the two primary alloying elements.
3. These copper alloys do not respond to heat treatment. The two digit suffix indicates minimum yield strength in every case.

There are four principal copper base alloy groups.

1. Pure copper is selected when high electrical or thermal conductivity is required. Special processing characteristics must be maintained to generate suitable conductivity and strength.
2. Bronze, typically made from mixtures of elemental copper and tin powders, is the most widely used nonferrous P/M material. The primary applications are oil impregnated self-lubricating bearings (see Section 3.2.2 Powder Materials for Bearing Applications). High

strength bronze structural components with tensile strengths as high as 345 MPa (50,000 psi) are usually made from prealloyed bronze powders. A typical designation for a low strength bronze alloy is CT-1000-13 indicating a nominal 10% tin content, no other alloying element and a minimum yield strength of 90 MPa (13,000 psi).

3. Brasses, in the form of prealloyed powders, are available in numerous compositions. Zinc content ranges from 10% to 30%. Leaded brasses contain copper contents from 70 to 90% and lead from 1 to 2%. Sintered brass components have tensile strengths as high as 275 MPa (40,000 psi). Machinability of the materials is comparable to cast and wrought brass stock of the same composition. Brass P/M components are well suited for applications requiring good corrosion resistance, good machinability, attractive appearance and ductility. A typical MPIF designation is CZP-3002-14 for a brass alloy with a nominal 30% zinc, 2% lead and a minimum 95 MPa (14,000 psi) yield strength.

   In an effort to eliminate lead because of its toxicity, brass powders containing bismuth and tin have been developed and are commercially available to replace machinable brass powders containing lead. The mechanical properties of the bismuth-tin-brass powders are better than those of leaded brass in some cases. They are only slightly inferior to their leaded brass counterparts in machinability and compressibility. Results of a recent study show that, if the proper machines, tools, and conditions are selected, the difficulty in machining regular non-leaded brass powder metal components (CuZn) is significantly reduced.

4. Nickel silver powders are prealloys of copper, nickel and zinc typically containing 16% to 19% nickel. Properties are similar to brass, but corrosion resistance is superior. A typical MPIF specification is CZN-1818-17 for a nickel silver alloy with a nominal 18% zinc, 18%, nickel and a minimum yield strength of 120 MPa (17,000 psi). The unleaded version is preferred because of the toxicity of lead.

3.2.1.5 Aluminum

Aluminum alloys have the widest application of the three lightweight metal groups (aluminum, beryllium and titanium) prevalent in the P/M industry. The major alloying elements are copper, magnesium and silicon. Tensile strengths range from about 105 to 280 MPa (15,000 to 40,000 psi) depending on composition, density and heat-treatment. Aluminum P/M offers the following advantages in structural components and bearings:

Good corrosion resistance
High strength-to-weight ratio
High rigidity-to-weight ratio
Excellent finishing properties
Excellent electrical and thermal conductivity
Very good dampening of sound and vibration

Similar to most P/M components, special processing is required to achieve the desired finishing characteristics. With such processing, P/M components can receive essentially the same finishing as equivalent wrought alloys. Where silicon is held to low limits, components can be anodized to achieve a variety of colors.

The Metal Powder Industries Federation has not yet established specifications for the various compositions of aluminum P/M. Commercial designations are sometimes used, and industry standards[8] have been adopted which enable the designer and P/M manufacturer to agree on critical factors such as chemical composition and minimum properties. Questions on characteristics are best resolved in discussions between the designer and part manufacturer.

3.2.2 POWDER MATERIALS FOR BEARING APPLICATIONS

P/M bearings can be made from most metal powders. Bronze is the predominant choice followed by ferrous alloys. The materials are fabricated to achieve 12 to 30% interconnecting microporosity, which functions as an oil reservoir. The high level of microporosity implies relatively low mechanical properties with tensile strengths typically ranging from 55 to 125 MPa (8,000 to 18,000 psi).

The MPIF system for designating bearing materials is identical to that for structural materials except that the three character suffix indicates radial crushing strength. The first suffix character is the letter K, followed by two numerals indicating the minimum value of the K factor in thousand pounds per square inch. The application of K factor to compute theoretical radial crushing force of a bearing is discussed in Section 2.4.1.

3.2.2.1 Bronze

Bronze containing a nominal 10% tin is essentially the only copper alloy used in bearings. The material offers excellent corrosion resistance and mechanical properties can be tailored to develop resistance to shock loading and ductility sufficient to allow staking. As density increases (microporosity decreases), the bearings become more ductile and support heavier loads. However, they retain less oil, restricting them to slower speed applications.

Graphite is added to improve lubricity. Materials with graphite content less than 0.3% are considered low graphite. Medium graphite bearing materials contain 0.5 to 1.8%. Materials with 3% to 5% graphite content are classified high graphite. Bronze bearings containing the higher graphite levels run very quietly, tend to require less field lubrication and operate at higher temperatures. A typical MPIF designation for a bronze bearing alloy is CTG-1004-K10 indicating a copper base with 10% tin, 4% graphite and a K factor of 10,000 psi.

3.2.2.2 Ferrous

Ferrous alloys are typically harder and stronger than 90-10 bronze. There are six alloys in this group: iron, iron-carbon, iron-copper, iron-copper-carbon, diluted bronze and iron-graphite. The effects of combining carbon and copper with iron are the same as for structural alloys, discussed in Section 3.2.1 Powder Materials-Structural Applications. Graphite imparts lubricity to ferrous alloys as it does with bronze. A typical MPIF designation for a ferrous bearing alloy is FC-0208-K25 indicating a ferrous based alloy with 2% copper, 0.6 to 0.9% combined carbon, and a K factor of 25,000 psi.

The term "diluted bronze" is applied to bronze that is diluted with 40 to 60% iron to lower the raw material cost. These bearings usually contain

---

8 See ASTM B 595

0.5% to 1.3% graphite for self lubrication. A typical MPIF designation is FCTG-3604-K22 indicating a ferrous based alloy with a nominal 36% copper, 4% tin, and a K factor of 22,000 psi. The graphite content of this alloy, which is not indicated in the designation code, is 0.5 to 1.3%.

3.3 MIM FEEDSTOCKS

MIM feedstock is prepared by intimately mixing fine metal powders into a thermoplastic carrier at an elevated temperature. The resultant mixture is hardened by cooling to room temperature, then granulated to produce the feedstock. The carrier is carefully selected to exhibit two characteristics:

- It must exhibit the characteristics of a stable thermoplastic during the injection molding process.
- It must be freely driven out of the molded preform during debinding.

Particle sizes typically range from 0.5 to 20 μm and are spherical in shape, compared with the coarser powders of irregular shape used in P/M. (See Table 4-3 in Section 4.2 Metal Injection Molding.) Fine particle sizes are required because, during sintering, the rate of shrinkage and densification of the green body depends on the particle size; very fine powders minimize sintering time. However, very fine particle size tends to drive up the cost of producing and handling the powder. Therefore, current MIM research is being directed at methods for reducing sintering times while utilizing feedstocks with coarser particle sizes.

Feedstocks are currently being produced for a variety of ferrous alloys including carbon, copper, nickel, tool and stainless steels. Non-ferrous alloys, such as monels and nichromes are also being produced.

Porosity is very low and is not interconnected in the finished MIM component. The very high density, which is greater than 96%, gives the MIM product mechanical properties equivalent to those achieved by investment casting in similar alloys. Table 3-1 shows tensile and elongation properties of MIM steels in the as-sintered condition.

3.4 P/F MATERIALS

P/F powders are made from a variety of ferrous compositions, including carbon, nickel, molybdenum and stainless steels.

3.4.1 P/F Powders

P/F powders are similar to P/M in many respects, with additional requirements for very low levels of inclusion content. This requirement is related to the very high dynamic properties achieved in P/F, such as impact energy, endurance limit and fracture toughness. These properties require a very homogeneous material, free from porosity and inclusions. Inclusions have a lesser effect than microporosity in P/M.

Powder producers have developed the capability to make forging grade water atomized low alloy steel powders with extremely low inclusion levels and cross contamination with iron powders. Quality control is assured by using image analysis techniques on microstructures from forged laboratory samples.

Some alloying elements used in cast and wrought alloys are not used in P/F due to the effects of atomization and sintering. For example, chromium and manganese are not used to promote hardenability because they

Table 3-1 MIM Materials Properties*

Low Alloy Steels

MIM Material Properties

Material Designation (condition)	Minimum Values			Typical Values					
	Tensile Properties			Tensile Properties			Density	Hardness	
	Ultimate Strength	Yield Strength (0.2%)	Elonga-tion (in 1 in.)	Ultimate Strength	Yield Strength (0.2%)	Elonga-tion (in 1 in.)		Apparent (direct)	Micro (converted)
	10^3psi	10^3psi	%	10^3psi	10^3psi	%	g/cm^3	Rockwell	
MIM–4600 as-sintered	37	16	20.0	42	18	40	7.6	45 HRB	
MIM–4650 as-sintered	55	25	11.0	64	30	15	7.5	62 HRB	
MIM–4650 quenched & tempered	215	190	<1.0	240	215	2	7.5	48 HRC	55 HRC
MIM–2700 as-sintered	55	30	20.0	60	37	26	7.6	69 HRB	

1993–1994 Edition
Approved: 1992   Revised:

Stainless Steels

MIM Material Properties

Material Designation (condition)	Minimum Values			Typical Values					
	Tensile Properties			Tensile Properties			Density	Hardness	
	Ultimate Strength	Yield Strength (0.2%)	Elonga-tion (in 1 in.)	Ultimate Strength	Yield Strength (0.2%)	Elonga-tion (in 1 in.)		Apparent (direct)	Micro (converted)
	10^3psi	10^3psi	%	10^3psi	10^3psi	%	g/cm^3	Rockwell	
MIM–316L as-sintered	65	20	40.0	75	25	50	7.6	67 HRB	
MIM–Duplex as-sintered	68	26	33.0	78	33	43	7.6	84 HRB	
MIM–17-4 PH as-sintered	115	94	4.0	130	106	6	7.5	27 HRC	
MIM–17-4 PH solution treated & aged	155	140	4.0	172	158	6	7.5	33 HRC	42 HRC

1993–1994 Edition
Approved: 1992   Revised:

*Taken from MPIF Standard 35 Material Standards for Metal Injection Molded Parts, 1993-1994 Edition.

oxidize during atomization and are difficult to reduce during sintering. Nickel, copper and molybdenum, which are more costly, are used because their oxides are reduced at conventional sintering temperatures.

Manganese sulfide (MnS) is sometimes added to P/F powders to increase machinability. The particles act as chip breakers and lubricants. In one test the machinability, based on the number of holes that could be drilled without resharpening, increased threefold with the addition of MnS. Since the MnS particles are added to the powder, there is very little diffusion of either the manganese or sulphur into the steel, as is the case with resulfurized steels that are used for conventional forged alloys. Therefore the modest reduction in some mechanical properties, such as tensile strength and ductility, is much less than with resulfurized steels.

The agreed designation for P/F powders consists of the prefix P/F followed by the standard ASTM four-or-five character designation. For example, P/F-1040 is a powder with a chemical composition similar to 1040 steel, reduced to powder, with additives such as lubricants to facilitate processing.

3.4.2 Mechanical Properties of Powder Forgings

Mechanical properties of powder forgings are dependent on process variables as well as the chemical composition of the material and the heat treatment. They may also vary with the direction of forging. Table 3-2 summarizes mechanical properties of powder forgings for several alloys and heat treatments.

3.5 ENGINEERING PROPERTIES OF P/M ALLOYS

The engineering properties of P/M alloys are similar in nature to those of wrought and cast. However, the inherent microporosity modifies the properties of P/M alloys. The properties are discussed in detail in MPIF Standard 35. This section gives a brief overview.

3.5.1 Density and Related Properties

There are three properties in this category: density, microporosity and permeability.

TABLE 3-2 TYPICAL MECHANICAL PROPERTIES OF P/F FERROUS STRUCTURAL COMPONENTS (ASTM B848)[1]

Alloy	Heat Treat[3]	Hardness (Rockwell)	Tensile str. MPa	Tensile str. 10³ psi	Yield str. MPa	Yield str. 10³ psi	Elong. %	Impact[2] (ft lbf)
P/F-1040	N	80 HRB	515	75	310	45	27	4
P/F-1040	Q	30 HRC	965	140	825	120	12	15
P/F-1060	N	80 HRB	585	85	345	50	22	2
P/F-1060	Q	40 HRC	1345	195	1205	175	8	10
P/F-10C40	N	97 HRB	690	100	480	70	15	3
P/F-10C60	N	23 HRC	790	115	670	100	11	2
P/F-11C60	N	28 HRC	895	130	620	90	11	3
P/F-4220	N	84 HRB	515	75	380	55	25	25
P/F-4240	N	93 HRB	620	90	415	60	18	12
P/F-4260	N	22 HRC	760	110	515	75	15	5
P/F-4620	N	96 HRB	550	80	415	60	20	25
P/F-4620	Q	28 HRC	965	140	895	130	24	60
P/F-4620	Q	38 HRC	1310	190	1070	155	20	35

1 Data do not constitute a part of the specification. They are based on specimens machined from sample blanks of the chemical composition, density and heat treatment specified.
2 Charpy V-notch impact energy.
3 N=Normalized, Q=Quench-hardened and tempered to the hardness listed.

Density

Most P/M component properties are closely related to final density, which is the weight per unit volume of the component expressed in grams per cubic centimeter (g/cm^3). Mechanical and structural component density is normally reported on a dry unimpregnated basis; density of bearings is reported on a fully oil impregnated basis. Density is calculated using the method given in MPIF Standard 42.

Density is also expressed as percent dense, which is the ratio of the density of a P/M component to that of a pore free material of the same chemistry and microstructure. In general, structural and mechanical components made by P/M have densities ranging from 80% to 95%. Many self-lubricating bearings have densities in the order of 75%, and filters usually have densities of 50%. Structural and mechanical components made by MIM and P/F typically have average densities up to 99.5%, with some areas virtually pore free.

Microporosity

Microporosity is the volume of voids expressed as a percentage of the total volume of the component. A component which is 85% dense has 15% microporosity. Microporosity can be present in two forms: an interconnected network that extends to the surface like a sponge, or a number of closed pores within the matrix. Interconnected microporosity is important to infiltration and the performance of self-lubricating bearings and is part of the specification for these materials. At very high densities, typically above 95%, particularly those achieved in MIM and P/F, there is less interconnection, and the microporosity becomes closed pore.

Microporosity, a unique structural characteristic of P/M components, is a controllable function of the raw material and processing techniques. Components can be produced either with nearly uniform microporosity or with variations (and density) from one section to another to provide different desired properties. For example, gears can be made self-lubricating in one area and dense and strong in other areas. The method for calculating pore volume or oil content of self-lubricating P/M components in terms of interconnected microporosity is given in MPIF Standard 35.

As microporosity is reduced to very low levels, typically below 1%, dynamic properties such as endurance limit and impact energy increase significantly. Microporosity forms a network of stress raisers; when they are eliminated, dynamic properties are improved.

Sound Damping

The porous nature of P/M components provides good sound damping. Ringing, common with wrought steel gears and other components, is reduced due to the controllable density in P/M components. This is an important benefit in products such as business machines where quiet operation is highly desirable. Damping decreases with increasing material density; it is aided by impregnation with sound damping materials such as polyesters. The controllable density of P/M components is also used to dissipate and muffle noise of air driven power tools.

Permeability

Another unique property that can be designed into P/M products is the ability to pass fluids or gas, as in filters. A P/M component can provide permeability ranging from highly restricted to highly open flow, depending on the forming and sintering techniques. The component can be produced with permeabilities that will separate materials selectively, diffuse the flow of gases or liquids, regulate flow or pressure drop in supply lines, or arrest flame by cooling gasses below combustion temperatures. Filters and flame arresters can be produced in almost any configuration, including sheets and tubes.

3.5.2 Mechanical Properties

Chemical composition of P/M components, as with wrought and cast metals, strongly influences the mechanical properties. However, in P/M components the properties depend on additional factors including: density; particle size; pore size, shape and distribution; and extent of sintering. Therefore, mechanical property data are commonly given in graphs showing the relationship between the property and density (or relative density). Figures 3-4 through 3-6 represent typical values and trends of properties versus density, based on data from test specimens conforming to chemical composition specified in MPIF Standard 35.

Test data obtained on standard specimens do not necessarily represent the actual component performance, because variations of density in the component may not be similar to those in the test specimen. Therefore test specimen data should be considered a guideline for component performance.

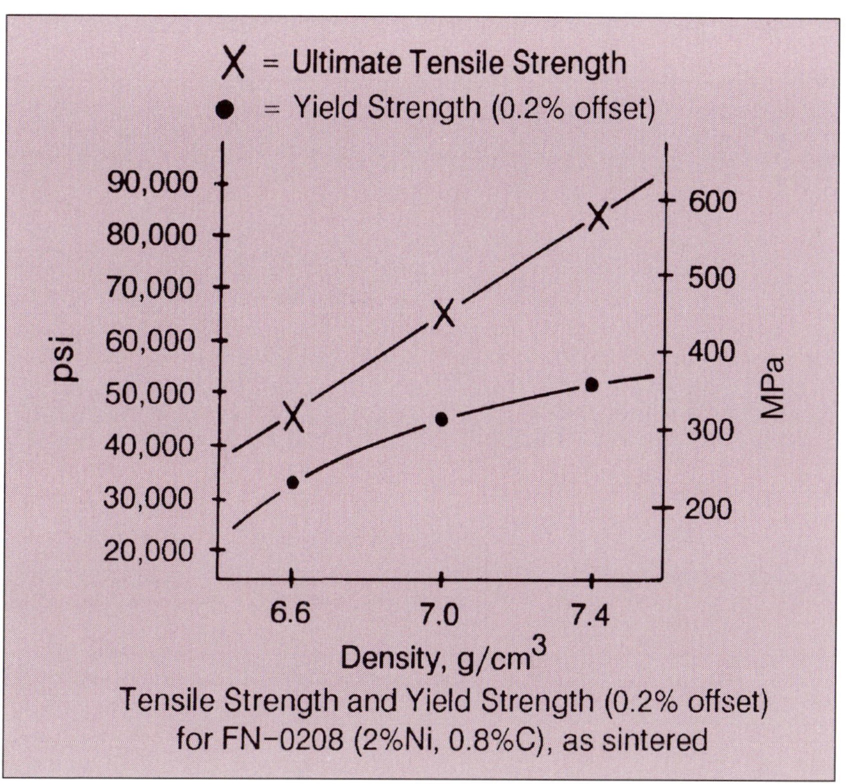

Figure 3-4 Tensile and yield strength versus density.

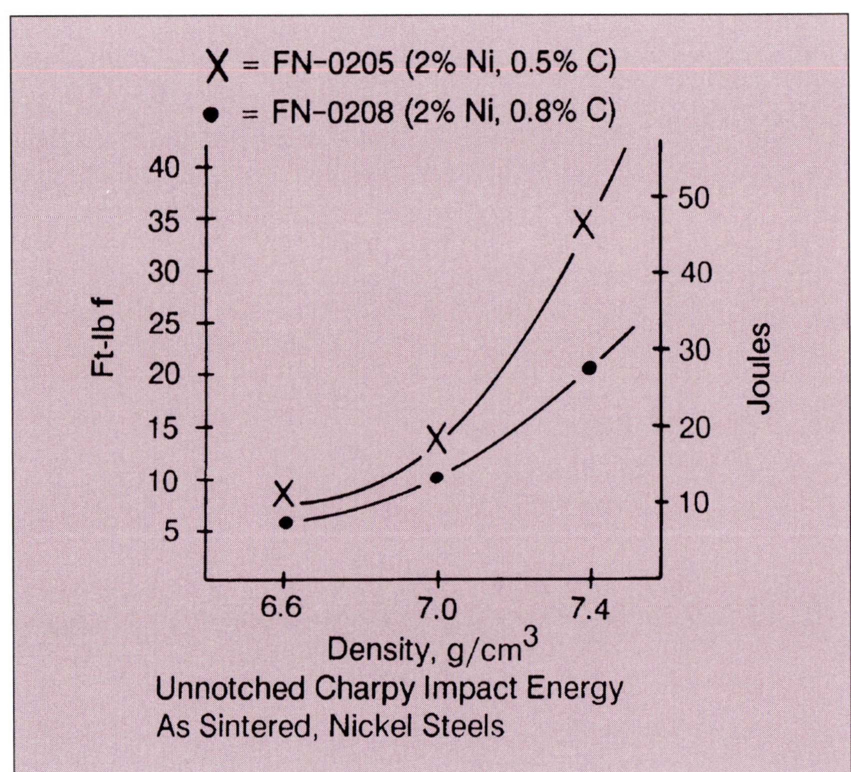

Figure 3-5 Charpy impact energy of nickel steels at various densities.

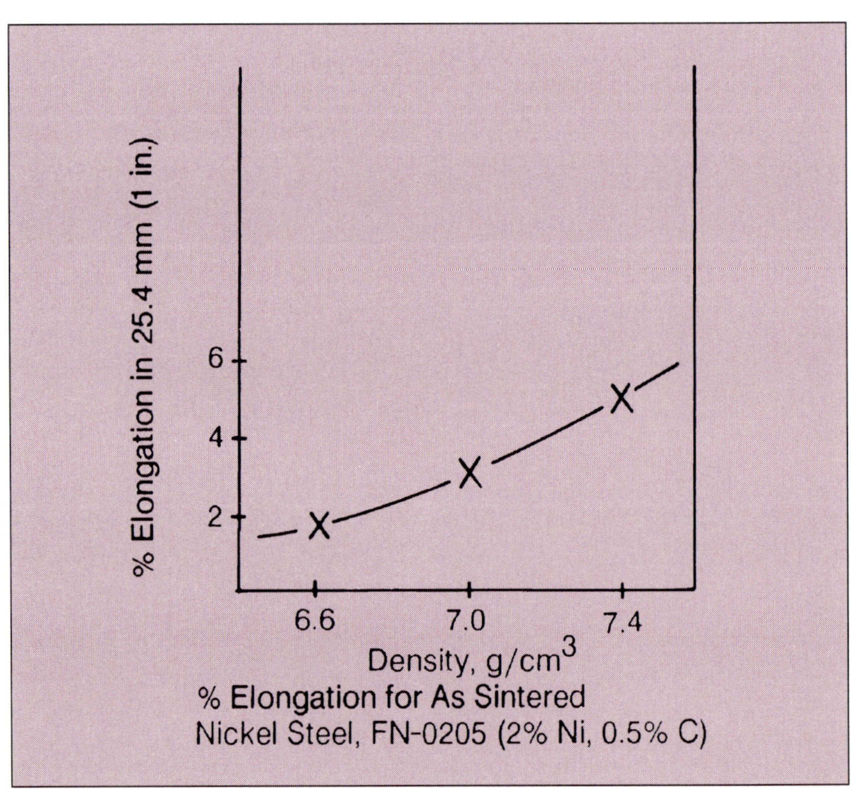

Figure 3-6 Elongation of nickel steel at various densities.

Strength

Figure 3-4 shows ultimate tensile and yield strengths of a 2% nickel and 0.8% carbon, pressed and sintered P/M material as a function of density.

Yield strength is generally closer to tensile strength than with wrought metals.

Figure 3-5 shows impact energy of a typical P/M nickel steel as a function of density, with impact energy rising significantly at higher densities and lower carbon contents.

Ductility

Ductility, the amount of plastic deformation prior to tensile fracture, is relatively low in P/M materials primarily due to microporosity. Ductility as a function of density is illustrated in Figure 3-6. Elongation is generally less than 10% for ferrous materials. However, elongations for some P/M brasses and stainless steel are as high as 15 to 20%. Ductility of most P/M materials can be increased considerably by hot repressing, or repressing followed by resintering or by high temperature sintering.

Hardness (Apparent)

Gross indentation hardness values of wrought metals and P/M components cannot be compared directly because of differences in structure. Hardness value of a P/M component is referred to as "apparent hardness," and is obtained using a standard tester and scale.[9] Apparent hardness is a combination of particle hardness and microporosity. Figure 3-7 shows how an indenter can compress the surface into underlying microporosity. However, carefully performed microhardness tests, such as Knoop or Vickers, will measure true particle hardness.

The area(s) in which hardness measurements are made should be agreed upon by the P/M producer and the user, and clearly specified because of possible density variation. Surface hardness values of case hardened components should be carefully interpreted because the indentations sometimes penetrate the case.[10]

3.5.3 Other Properties

Corrosion Resistance

Microporosity in P/M components significantly affects corrosion resistance because it may allow corrosive media to be trapped and retained. Higher density improves corrosion resistance, as it does most

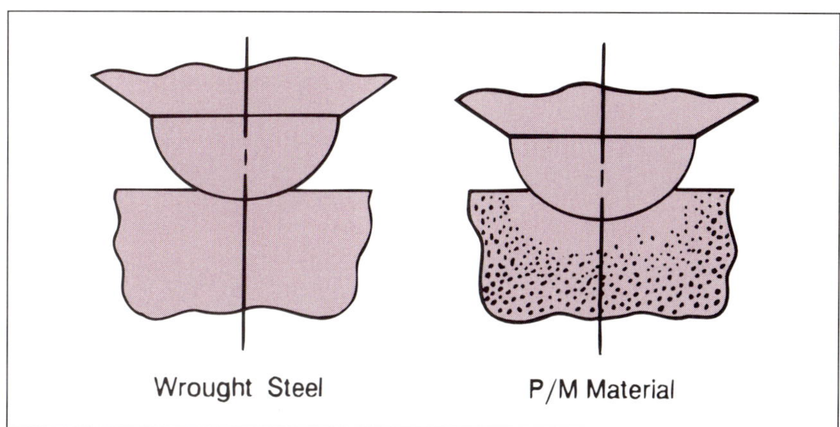

Figure 3-7 Hardness indenter penetrates deeper in P/M material because of lower density.

9 See MPIF Standard 43
10 See MPIF Standards 51 and 52

other properties. Stainless steel P/M components have relatively good corrosion resistance in the atmosphere and in weak acids. Careful control of the atmosphere dew point and cooling rate permit the production of stainless steel parts meeting or exceeding the 100-hour 5% NaCl test without any rust.

Surface Finish

Excellent surface finish is an inherent feature of P/M components. The overall smoothness and surface reflectivity depend on density, tool finish and secondary operations. Conventional profilometer readings give an erroneous impression of surface smoothness (except at near pore free levels) in that a different surface condition exists than is the case with machined or ground surfaces. Conventional readings (RMS) take into account the peaks and valleys of machined surfaces, while P/M components have a series of very smooth surfaces which are interrupted with varying size pores. Because of the microporosity, conventional RMS readings can be misleading, generally showing a higher reading than that obtained with comparison methods.

Effective surface smoothness of P/M components compare favorably with ground or ground-and-polished surfaces of wrought and cast components. Surface smoothness can be further improved by secondary operations such as repressing, honing, burnishing or grinding. The surface finish requirements and method of determination must be established by the purchaser and producer to meet the design intent. Where very high densities are attained, usually by liquid phase sintering, MIM or P/F, surface smoothness approaches that of wrought or cast products made from equivalent alloys.

3.5.4 Summary

Properties of P/M materials are documented in MPIF Standard 35. The summary in Tables 3-3 through 3-5 will serve as an overview to guide the designer to the appropriate alloy group for a wide variety of applications.

TABLE 3-3 PROPERTIES OF FERROUS STRUCTURAL MATERIALS

Material	Density g/cm^3	Tensile Strength[a] MPa	psi	Comments
Carbon steel				
MPIF F-0008	6.9	370[b]	54,000[b]	Often cost effective
0.8% c.c.[d]	6.9	585[c]	85,000[c]	
Copper steel				
MPIF FC-0208	6.7	415[b]	60,000[b]	Good sintered strength
2% Cu, 0/8% c.c.	6.7	585[c]	85,000[c]	
Nickel steel				
MPIF FN-0205	6.9	345[b]	50,000[b]	Good heat treated
2% Ni, 0.5% c.c.	6.9	825[c]	120,000[c]	strength, impact energy
MPIF FN-0405	6.9	385[b]	56,000[b]	
4% Ni, 0.5% c.c.	6.9	845[c]	123,000[c]	
Infiltrated steel				
MPIF FX-1005	7.3	530[b]	77,000[b]	Good strength, closed off
10% Cu, 0.5% c.c.	7.3	825[c]	120,000[c]	internal porosity
MPIF FX-2008	7.3	550[b]	80,000[b]	
20% Cu, 0.8% c.c.	7.3	690[c]	100,000[c]	
Low alloy steel, prealloyed Ni, Mo, Mn				
MPIF FL-4605 HT	6.95	895[c]	130,000[c]	Good hardenability, consistency in heat treatment
Stainless steels				
MPIF SS-316 N2 316 stainless	6.5	415[b]	60,000[b]	Good corrosion resistance, appearance
MPIF SS 410 HT 410 stainless	6.5	725[c]	105,000[c]	

a Reference: MPIF Standard 35, Materials Standards for P/M Structural Parts. Strength and density given as typical values.
b As-sintered
c Heat treated
d Combined carbon

TABLE 3-4 SOFT MAGNETIC COMPONENTS

Material	Density g/cm^3	Maximum Magnetic Induction (kG)[a]	Coercive Field (Oe)
Iron, low density	6.6	10.0	2.0
Iron, high density	7.2	12.5	1.7
Phosphorous iron, 0.45% P	7.0	12.0	1.5
Silicon Iron, 3% Si	7.0	11.0	1.2
Nickel iron, 50% Ni	7.0	11.0	0.3
410 Stainless Steel	7.1	10.0	2.0
430 Stainless Steel	7.1	10.0	2.0

a Magnetic field -15 oersteds

TABLE 3-5 NONFERROUS COMPONENTS

Material	Usage and Characteristics
Copper	Electrical components

Bronze
    MPIF CTG-1001
    10% tin, 1% graphite
	Self-lubricating bearings:
	@ 6.6 g/cm³ oiled density, 160 MPa (23,000 psi)
	K strength[a], 17% oil content by volume

Brass
 	
    MPIF CZ-1000
    10% zinc
    MPIF CZ-3000
    30% zinc
	Electrical components, applications requiring
	good corrosion resistance, appearance and ductility
	@ 7.9 g/cm³, yield strength 75 MPa (11,000 psi)
	
	@ 7.9 g/cm³, yield strength 125 MPa (18,300 psi)

Nickel silver
    MPIF CNZ-1818
    18% nickel, 18% zinc
	Improved corrosion resistance, toughness
	@ 7.9 g/cm³, yield strength 140 MPa (20,000 psi)

Aluminum alloy	Good corrosion resistance, light weight, good electrical and thermal conductivity
Titanium	Good strength/mass ratio, corrosion resistance

a Radial crushing constant. See MPIF Standard 35, Materials Standards for P/M Self-Lubricating Bearings

Manufacturing Processes

There are five basic manufacturing processes that utilize metal powders. They are conventional powder metallurgy (P/M), which is the dominant sector of the industry, metal injection molding (MIM), powder forging (P/F), hot isostatic pressing (HIP) and cold isostatic pressing (CIP). This section describes the first three, which are most commonly specified. Further information on the other two can be obtained from the Metal Powder Industries Federation.

4.1 POWDER METALLURGY (P/M)

There are three basic phases in the P/M process: powder preparation, compaction and sintering. Powder preparation is discussed in Section 3. Compaction compresses blended powder into net or near net shape green compacts with sufficient strength for handling. Sintering develops mechanical properties. Secondary operations, when required, can further increase mechanical properties, improve dimensional tolerances, generate features not formed in compaction, and apply surface finishes for corrosion protection and appearance.

4.1.1 COMPACTION[11]

The mechanics of the compaction cycle largely determine the shapes that are practical to produce by the P/M process.

4.1.1.1 Cycle Events

Conventional compaction is performed in hard tooling in a cycle consisting of four events shown in Figure 4-1. The volume of powder fed into the die is controlled by the position of the lower punch relative to the die when the die is filled. After filling, the feed shoe is withdrawn and the upper punch enters the die. Different types of compacting systems are

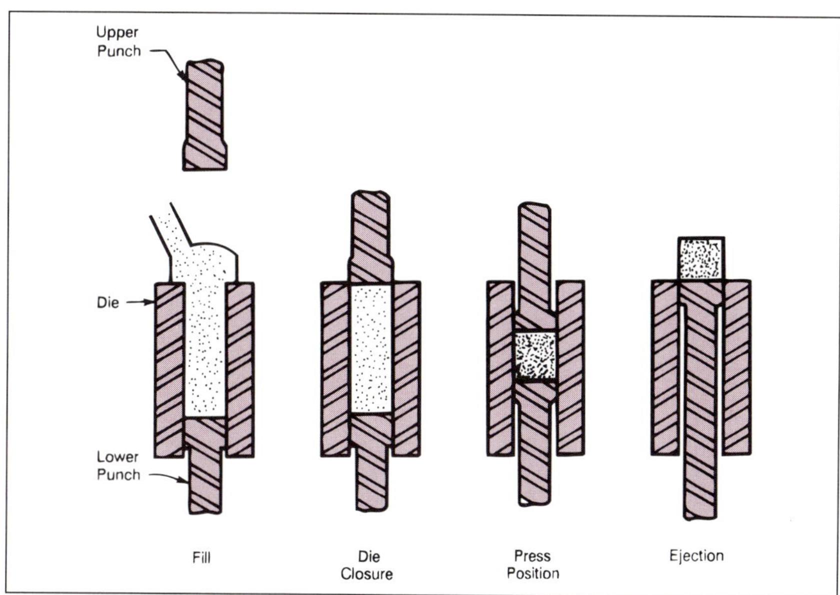

Figure 4-1 Typical tool motions during a powder compaction cycle.

---

11 Adopted from German, R.M. Powder Metallurgy Science, Second Edition, Chapter 6, Metal Powder Industries Federation, Princeton, NJ, 1994

available for controlling the relative motion of the tool members. In all cases, force is applied to each end of the compact through the punches. When compaction is completed, the upper punch withdraws from the die and relative motion between the lower punch and die ejects the compact. Compaction cycles are classified as single, double or multiple action. Conventional press production rates can vary from 200 to 2000 pieces per hour and are dependent on material filling properties, component geometry and size, and machine type.

Movement of only one of the prime tool elements, usually the upper punch, is termed single action. Relative motion of two members, either both punches with die stationary as shown in Figure 4-1 or upper punch and die with lower punch stationary, is termed double action. Multiple action utilizes several tool members so arranged and controlled that a separate member supports each level of the component. This type of compaction is often used for class 4 components (defined below).

The behavior of the powder during compaction significantly affects the properties of the P/M component. As compaction begins, the powder particles reorient to a more dense particle packing level with progressively reduced microporosity, and the area of surface contact between particles increases. As compaction progresses, microporosity is further reduced by continued realignment and by plastic deformation of the particles. These events are illustrated in Figure 4-2. The fill ratio, which is the green component (as pressed) density divided by the apparent density of the material (powder density at fill) is typically between 2:1 and 2.5:1.

The final compaction pressure, which is developed at the end of the stroke, ranges from 110 to 275 MPa (8 to 20 tons per square inch) for copper and aluminum powders to a maximum of 690 MPa (60 tons per square inch) for ferrous powders. The high pressures cause cold welding and interlocking of the particles, imparting sufficient strength to the compact, so that it can be ejected from the die and handled for sintering. The strength of the compact after pressing and before sintering is termed the green strength.

The extremely high pressure causes elastic deformation in the compact.

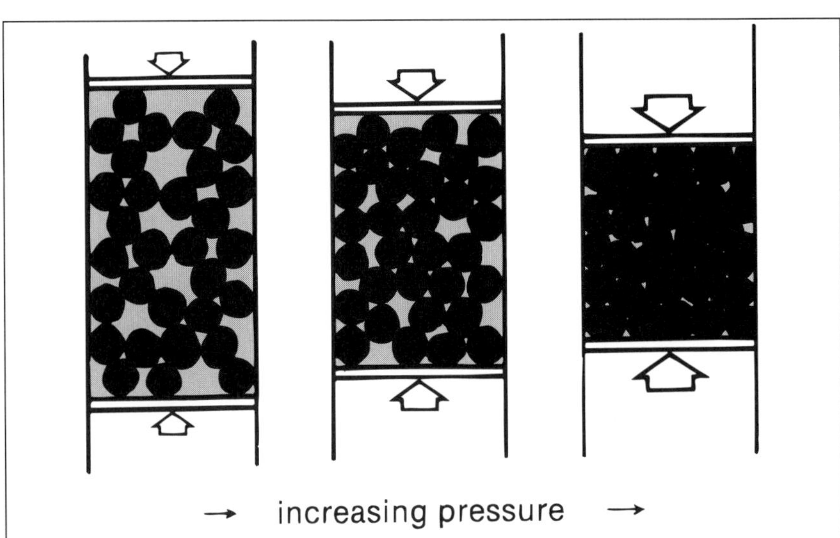

Figure 4-2 A simplified view of the stages of metal powder compaction. Initially, repacking occurs with the elimination of particle bridges. With higher compaction pressures, particle deformation is the dominant mode of densification.[12]

12 Ibid p. 206

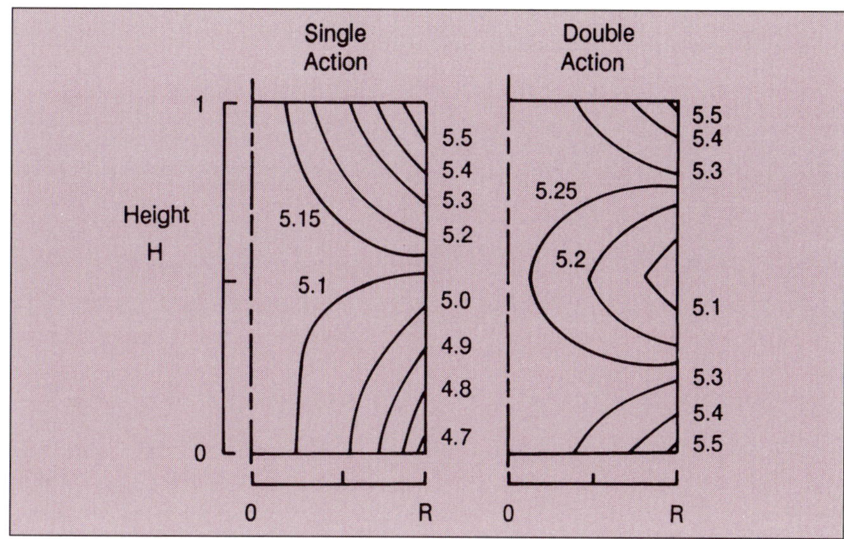

Figure 4-3 Constant density lines in cylinders of compacted copper powder. The contours are in g/cm³ and are plotted versus height and radius (from the centerline axis) for both single and double action pressing (based on Thompson).[13]

When the pressure is released on ejection, the elastic deformation is recovered (termed "springback") so that the compact dimension exceeds the die dimension. The springback is usually 0.3% or less of the die dimension.

The mechanics of powder compaction are largely governed by friction between the die and the powder and between powder particles. Friction losses cause a reduction in compaction pressure, and consequently in the density of the compact. Adding the correct amount of lubricant to the powder blend reduces friction between the powder particles and tool members. Double action compaction reduces density variations parallel to punch movement. Figure 4-3 shows typical powder density distribution in cylinders of copper powder compacted by single and double action.

Density gradients within a P/M component are usually undesirable because of material property reductions and nonuniform dimensional changes during sintering. Density gradients and controlled microporosity levels are deliberately developed in some cases to gain product advantages. For example, one section may require lower density to allow for impregnation while another requires maximum density for maximum strength.

Density gradients impose a limit on the length of the compact. The most important parameter for determining the severity of density gradients and maximum allowable length is the ratio of length to diameter (l/d). When l/d is low, single action compaction may be adequate. Successful compaction is not generally achieved when l/d exceeds 5:1. Thin wall sections are often difficult to fill properly, causing slower production rates and reduced part consistency. Where the ratio of length to wall thicknesses is 8:1 or greater, special precautions must be taken to achieve uniform fill, and variations in density are virtually unavoidable.

Double action compaction generates a plane perpendicular to the die axis between the punches where there is essentially no axial movement of powder. A line in this plane on full section views of P/M component is termed a "neutral axis" (See Figure 4-4). Placement of the plane can be varied by controlling punch movements during compaction. There is

13 Ibid p. 219

some latitude to locate the plane to achieve favorable density gradients. In some cases, the location of the plane is dictated by the configuration of the compact, such as a change in level.

4.1.1.2 Classes of Components and Levels of Features

P/M components can be classified into one of four classes according to their complexity and difficulty of manufacture. Class 1 are single level components with simple shapes and a low l/d that allow single action compaction (Figure 4-5). Class 2 single level with higher l/d requiring double action (Figure 4-6). Class 3 have two levels, and are usually termed external and internal flanged components (Figure 4-7). Class 4 have three

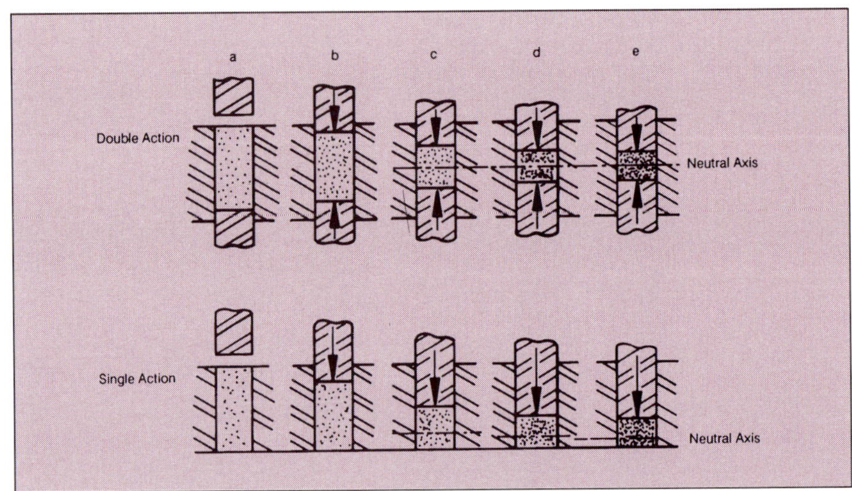

Figure 4-4 Double acting (above) and single acting (below) compaction of metal powders. Note the symmetrical pressure propagation within the powder mass that results from double action compaction.

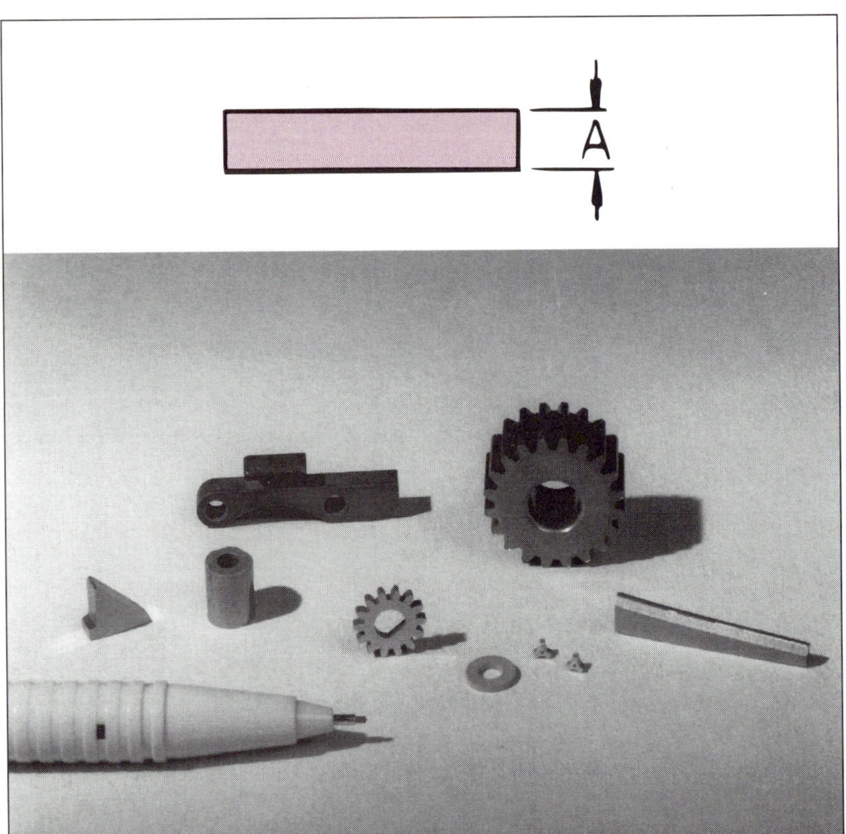

Figure 4-5 Class 1 components. Class 1 parts are thin, one-level parts of any contour that can be pressed with a force from one direction. The A dimension is generally limited to a maximum of 6.35 mm (0.20 in).

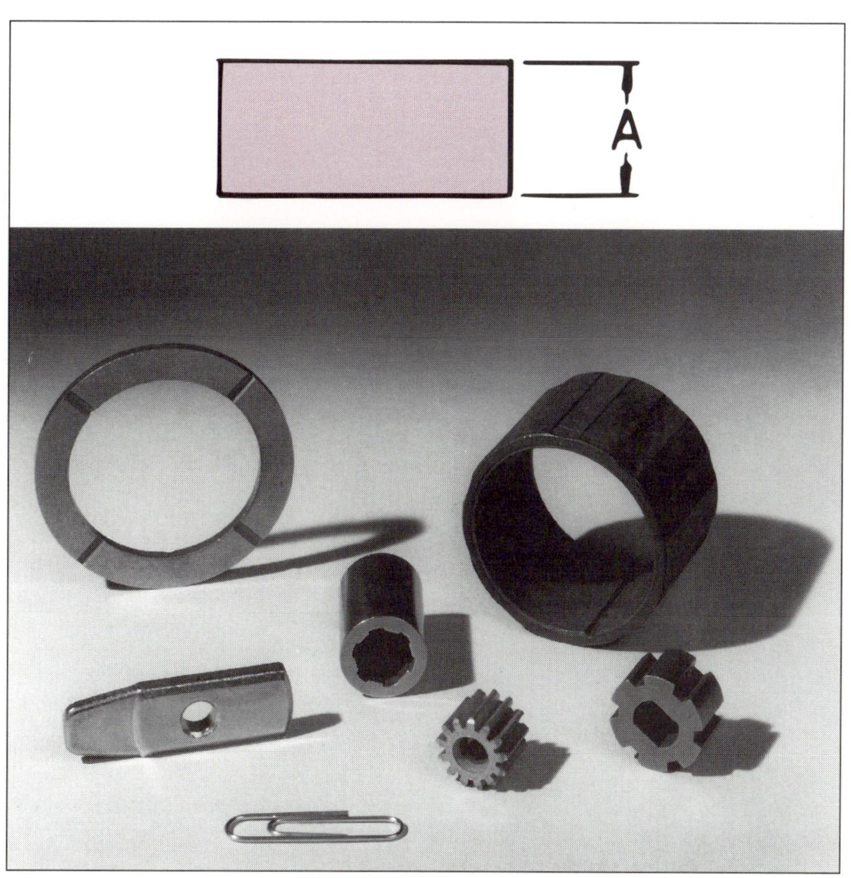

Figure 4-6 Class 2 components. Class 2 parts are one-level parts of any thickness and contour that must be pressed with forces from two directions.

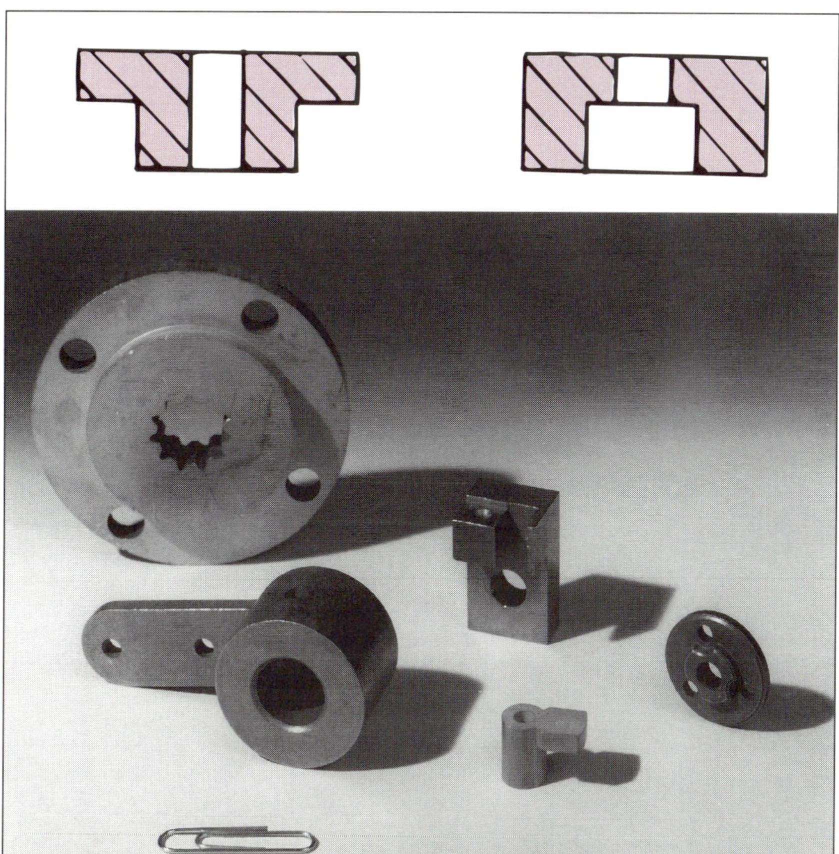

Figure 4-7 Class 3 components. Class 3 parts are two-level parts of any thickness and contour that must be pressed with forces from two directions.

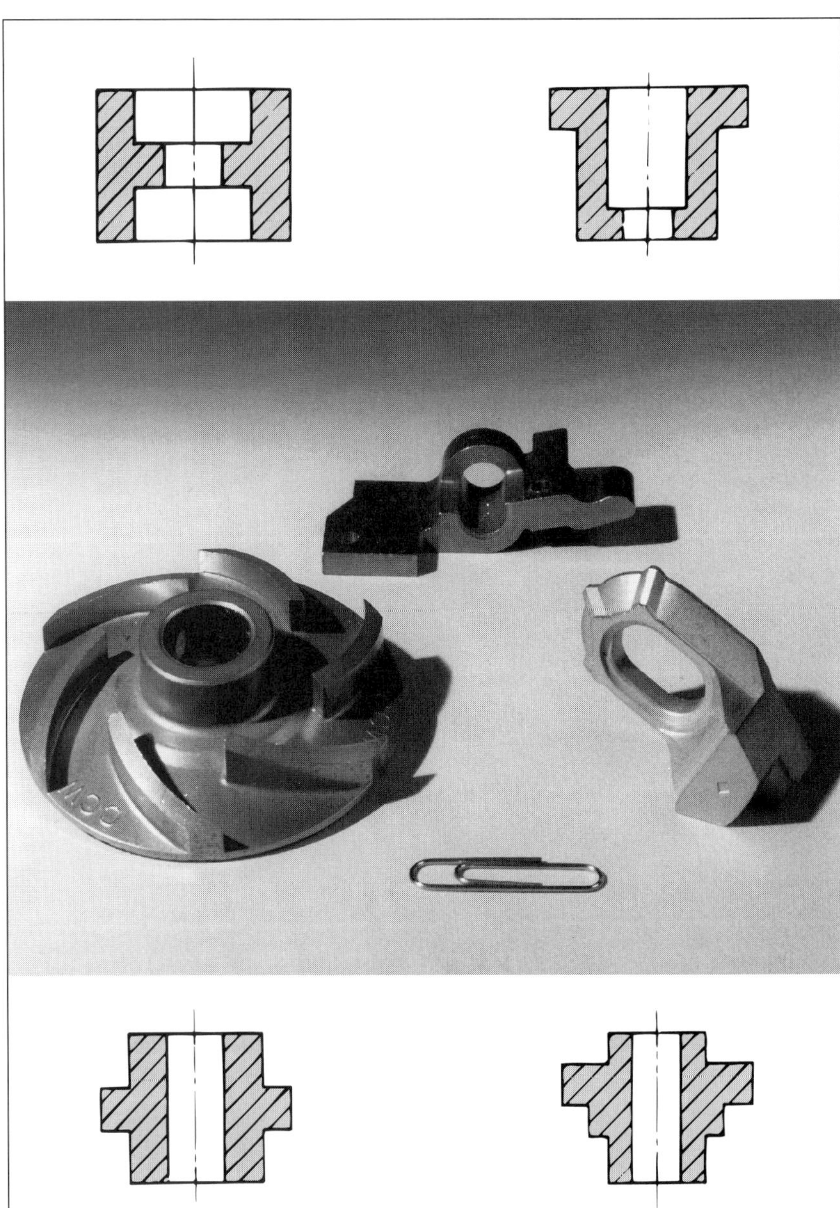

Figure 4-8 Class 4 components. Class 4 parts are multi-level parts of any thickness and contour that must be pressed with multiple forces from two directions.

or more levels (Figure 4-8). Class 3 and 4 components require double action compaction because of their multiple levels. Multiple action compaction is required for Class 4 components to minimize density gradients. The four classifications are summarized in Table 4-1. Because tooling complexity increases for each level, manufacturing costs can be contained by designing to the lowest class that meets the application requirements.

	TABLE 4-1 P/M COMPONENTS CLASSIFICATION	
Class	Levels	Press Action
1	1	Single
2	1	Double
3	2	Double
4	more than 2	Double and Multiple

4.1.2 SINTERING

Sintering is the process whereby metal powder particles bond at temperatures below the melting point of the major constituent by atomic transport events. Movement of the metal, on an atomic scale, reduces the net surface energy by reducing surface area. The contact points between particles, which were formed during compaction, increase in size and strength, improving mechanical properties. The compact can additionally shrink, grow, increase in conductivity or become harder.

Shrinkage and growth are defined as the change in dimension from the die dimension to the sintered dimension. Shrinkage is opposite in effect to the increase in dimension due to springback that occurs on ejection from the die; growth is additive.

Shrinkage is primarily caused by the solid state diffusion, which increases the size of the contact areas between particles, reducing pore volume and therefore increasing density. The amount of shrinkage varies inversely with the green density. Therefore large density gradients result in nonuniform shrinkage, which can lead to conditions such as nonuniform dimensional changes, distortion, residual stresses and cracks.

4.1.2.1 The Sintering Operation

In the typical sintering operation, the green compacts are placed either manually or automatically on a conveyer that travels through the furnace at a controlled rate depending on the alloy being sintered. Small and medium size components are generally transported on conveyers using alloy belts 305 to 915 mm (12 to 36 inches) wide. Roller hearth furnaces are used for larger components that are carried in alloy trays on driven rollers. Walking beam, pusher and vacuum furnaces are used where temperatures exceed the capabilities of alloy belts or roller hearth furnaces. Ceramic or graphite plates keep components flat during sintering. Thin flat components are stacked on the plates to maximize furnace throughput.

Typical sintering temperature ranges are 790 to 845°C (1450 to 1550°F) for bronze alloys and 1095 to 1150°C (2000 to 2100°F) for ferrous alloys. High temperature sintering at temperatures up to 1315°C (2400°F) is being utilized to improve properties of ferrous components. There are usually three zones in a typical furnace: preheat, high heat and cooling. The preheat zone removes lubricants and other organic components of the powder mix, and raises the temperature of the compact. Sintering is accomplished in the high heat zone. The components are cooled in a protective atmosphere below the oxidizing range in the cooling zone.

Belt furnace cycles average 45 to 75 minutes for small bronze bushings and two to three hours for average size ferrous components. A typical furnace temperature cycle is illustrated in Figure 4-9. Throughput varies with component size, belt width and belt speed, reaching as great as 450 kg/hr (1000 lb/hr) in large furnaces.

High Temperature Sintering

As sintering temperatures for ferrous alloys are increased to the higher end of the sintering range of 1315°C (2400°F), the mechanical properties are generally improved. The improvement comes primarily from more complete interparticle bonding and improved diffusion of alloying elements. Density is also increased, and this effect contributes particularly to

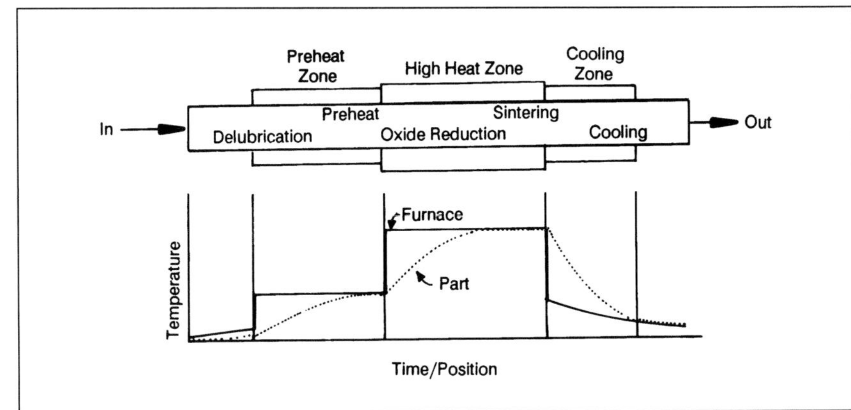

Figure 4-9 The sequence of operations occurring in a typical sintering furnace. The lower diagram shows the time-temperature profile typical to metal powder sintering.[14]

TABLE 4-2 EFFECTS OF SINTERING TEMPERATURE ON MECHANICAL PROPERTIES OF TWO GRADES OF NICKEL STEEL[15]

MPIF Alloy	FN-0205	FN-0205	FN-0208	FN-0208
Sintering Temperature	1120°C	1260°C	1120°C	1260°C
	2050°F	2300°F	2050°F	2300°F
Tensile strength				
MPa (psi)	380 (55,000)	550 (80,000)	450 (65,000)	760 (110,000)
Yield strength				
MPa (psi)	190 (28,000)	415 (60,000)	330 (48,000)	585 (85,000)
% elongation	4	7	2	4
Impact energy				
J (ft-lbf)	19 (14)	38 (28)	11 (8)	33 (24)
Hardness, HRB	64	80	80	90
Density, g/cm³	7.20*	7.32	7.20*	7.30

* The density was not recorded, but was estimated based on the green density (before sintering), which was 7.20 for all samples.

dynamic properties such as impact energy and endurance limit. The effects of densification on dynamic properties are most pronounced when porosity is reduced to 0.5% or less.

Table 4-2 compares the properties of two nickel steels, FN-0205 and FN-0208, for samples sintered at 1120°C (2050°F) and 1260°C (2300°F). The sintering atmosphere at the higher temperature was somewhat different from that at the lower temperature; other parameters were essentially identical.

Copper steels show similar, but less significant improvement. Copper melts at conventional sintering temperatures, producing a liquid phase, which accelerates sintering and improves mechanical properties. The increased sintering temperature brings about a further improvement.

Prealloyed austenitic stainless steels (300 series) respond in a different way to high temperature sintering. In one test that compared sintering temperatures of 1120°C (2050°F) with 1260°C (2300°F), tensile and yield strengths were nearly equal, but elongation more than tripled at the higher temperature. When nitrogen was introduced into the sintering atmosphere, strength increased and the ductility was lowered due to absorption of nitrogen from the sintering atmosphere

---

14 Ibid p. 288
15 Sanderow, H., Ed, High Temperature Sintering, Metal Powder Industries Federation, Princeton, NJ, 1990

Liquid Phase Sintering

Liquid phase sintering is sometimes used to improve mechanical properties by reducing microporosity, achieving nearly a pore free condition. The process utilizes an additive to the powder mix, which liquifies at a temperature below the melting point of the base powder. The additive must be one that will wet the base powder, and one in which the base powder is soluble. Sintering cycle parameters are adjusted to achieve the desired results.

The liquid phase distinguishes this process from conventional sintering, in which no liquification occurs. The process is shown schematically in Figure 4-10. During the solution reprecipitation stage, the solid dissolves into the liquid, and the small particles dissolve and reprecipitate on the larger particles. The solid skeleton is formed by rearrangement of the larger particles.

The increased densification that occurs during liquid phase sintering causes shrinkage of the compact. The amount of shrinkage depends on the density of the compact. The shrinkage is predictable, and the compaction tools can be adjusted to provide for it, so that dimensional repeatability is not adversely affected.

Sinter-Hardening

P/M components may be heat treated by a number of operations as described in section 4.1.3.3. However, these are traditionally secondary

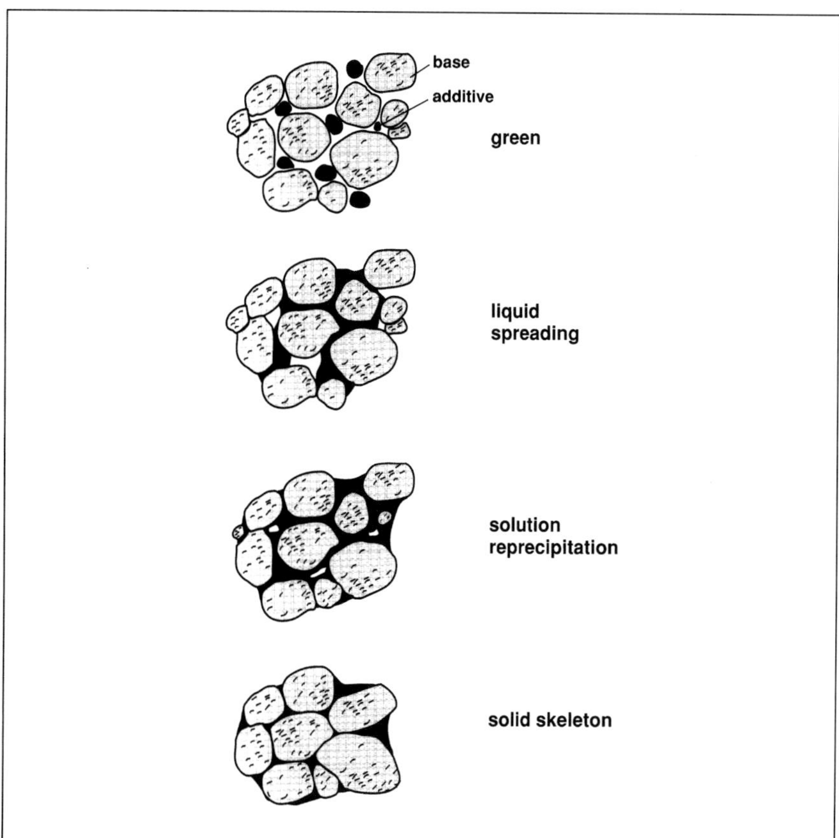

Figure 4-10 The conceptual stages to liquid phase sintering using a mixture of two powders. The base powders remains solid during the process and the additive powder is responsible for formation of the liquid.[16]

16 German, R.M., Powder Metallurgy Science, Second Edition, Chapter 7, Metal Powder Industries Federation, Princeton, NJ, 1994

operations, which entail processes such as reheating the components in a controlled atmosphere, followed by quench hardening and tempering. Under certain circumstances the need for secondary hardening processes may be eliminated.

The rate of cooling experienced by parts following sintering influences the microstructural constituents that are formed. The microstructural constituents present depend on the effective cooling rate, mass and shape of the part (ruling section) and material composition (hardenability).

Conventionally sintered copper and nickel steels, such as FC-0208 and FN-0205, usually yield microstructures consisting of ferrite and pearlite. More highly alloyed materials, such as partially alloyed low-alloy steels, may form some bainite in addition to ferrite and fine unresolved pearlite. However, if the sintering furnace is modified to permit the cooling rate to be increased, it is possible to produce martensite in the microstructure.

Sinter-hardening refers to a process where the cooling rate experienced in the cooling zone of the sintering furnace is fast enough that the steel matrix is transformed into bainite and/or martensite. Depending on the composition and the mass of the component, cooling rates of 1 to 10°C per second, through the temperature range from 900 to 425°C (1650 to 800°F), are generally necessary.

Whether sinter-hardened or quench-hardened, components should be tempered to relieve stress and improve toughness.

4.1.2.2 The Sintering Atmosphere

A controlled atmosphere is maintained inside the furnace during sintering for any of several reasons.

1. Help remove and "burn off" lubricants and other organic components from the metal powder mix.
2. Reduce surface oxides from the powder surfaces and enable diffusion to take place as an essential part of the sintering process.
3. Provide a uniform heat transfer to and from the compact through conduction and convection.
4. Prevent oxidation of the compact as it travels through the heating and cooling sections of the furnace.
5. Vary the alloy composition of the compact by adding or removing elements such as carbon and nitrogen.

Traditional sintering atmospheres include endothermic gas produced from natural gas and dissociated ammonia from anhydrous ammonia. More recently, atmosphere systems based on pure nitrogen and pure hydrogen from cryogenic sources are being used. Vacuum or partial pressure vacuum is also used for sintering of certain alloys in a batch type furnace. The dew point, hydrogen content, carbon potential and oxygen content of the sintering atmosphere are monitored and controlled to obtain the desired properties in the P/M component. The choice of the atmosphere type is based on factors such as the material being sintered, the furnace design, and the properties required of the P/M component as well as economic considerations.

4.1.2.3 Mixed Phase Sintering

Mixes of elemental powders are normally used for most applications relying on partial or complete alloying to occur during sintering. The process is termed "mixed phase sintering." Mixtures of elemental powders offer several advantages over prealloyed powder including:

1. Composition is easy to change.
2. Compaction pressures are lower because elemental powders have better compressibility.
3. Higher green density and strength are achieved.
4. It is possible to form unique microstructures.

Mixed phase sintering requires extra care in material selection and processing because:

1. The composition may not be fully homogenous; local variations may occur. The variations require both time and temperature control to ensure homogenization.
2. The process works best with fine particles where diffusion distances are small, but fine particles may cause compaction difficulties such as poor powder flow, tool wear and powder wedging between tool members.
3. Two powders with sufficiently different diffusion rates can cause swelling.
4. Detrimental phases, such as brittle intermetallics, can form with improper control of the sintering cycle.

4.1.2.4 Infiltration and Assembly

Infiltration and assembly are frequently accomplished during sintering. Infiltration is performed by positioning a preform of the infiltrant material, with a lower melting point than the primary metal, on or under the pressed green (presintered) compact. For example, copper and its alloys are the most common infiltrant for iron. The infiltrant melts before the sintering temperature is reached. Some is diffused into the powder particles of the primary metal (depending on solubility), and some is deposited in the pores of the compact by capillary action, forming a composite structure. Properties depend on the metals used, their proportions, and the conditions under which they were combined. In addition to improving mechanical properties, infiltration seals pores in preparation for electroplating, improves machinability, and essentially makes the components gas or liquid tight.

Assembly is accomplished by fitting the components together prior to sintering, as shown in Figure 4-11. Slots and keyways are used for location and orientation. Infiltration is used to effect a bond similar to brazing or soldering. Another method is to formulate one component for low growth (or shrinkage) and the other for high growth. This technique, known as sinter bonding, develops a bond that is superior to a press fit.

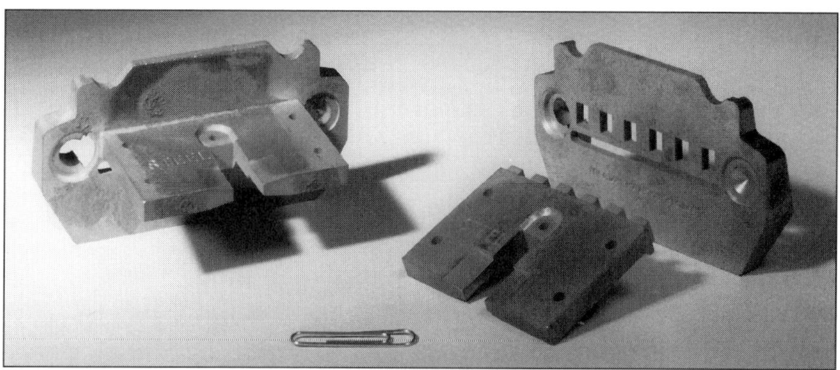

Figure 4-11 Assembly consisting of two P/M parts joined during sintering.

4.1.3 SECONDARY OPERATIONS

One of the major P/M advantages is its capability to produce components to net or near net shape with few or no secondary operations. Sometimes components may require closer tolerances, increased mechanical properties, features not possible by compaction, surface protection or enhanced appearance. Most of the operations that accomplish these features are performed on P/M components in the same manner as on cast or forged components. However, microporosity frequently imposes limitations on some secondary operations. Molten salts, cutting fluids, cleaning solutions and plating solutions can be trapped in the pores of the component causing corrosion. If fluid use is essential to the process, it must be completely removed or chemically compatible with the P/M alloy and application environment.

Six types of operations are commonly performed: repressing, impregnation, heat treating, machining, surface treatment and welding/brazing.

4.1.3.1 Restriking

Restriking usually requires different tooling than that employed for compaction. There are two types of restriking operations: repressing and sizing. Both cause plastic deformation, and both may deform metal closing surface microporosity, negating the advantages of the pores.

Repressing reduces volume thereby increasing density and improving strength, hardness and dimensional accuracy. Repressing can also be used to modify the surface shape, such as embossing (coining).

Sizing primarily improves dimensional accuracy with little or no increase in density or strength. It is accomplished by burnishing using die or core rod sizing tools. Reaming, broaching and boring are sometimes called sizing operations when they are used to bring an inside diameter to finished dimension. They are, strictly speaking, machining operations because they are accomplished by removing, rather than by displacing metal. Figure 4-12 shows the sequence of events in sizing a bushing.

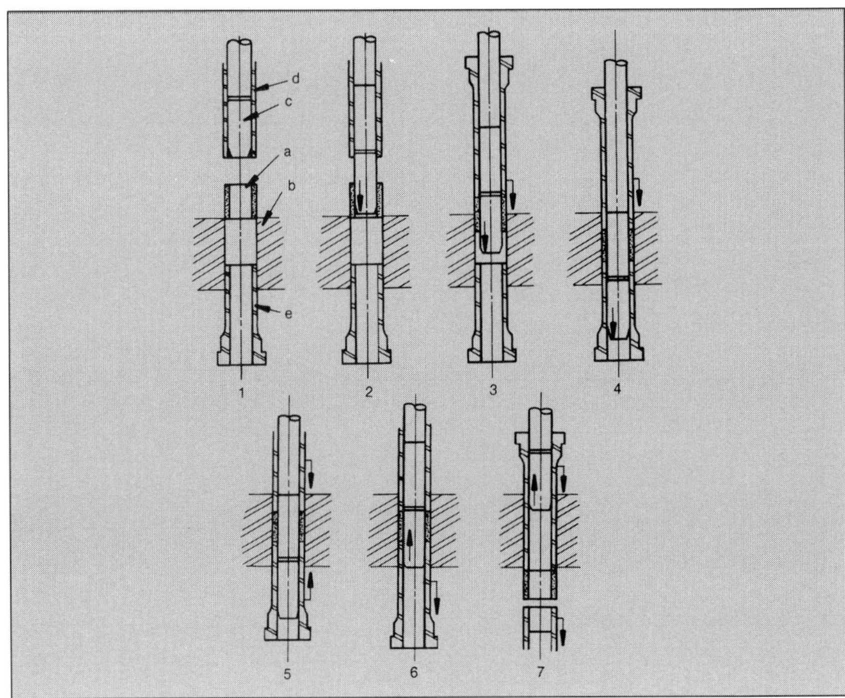

Figure 4-12 Sizing operation on a bushing:
a. Bushing b. Die c. Core rod d. Upper punch e. Lower punch.

4.1.3.2 Impregnating

Controlled microporosity permits P/M components to be impregnated with oils or resins. Oil impregnation is achieved by soaking the components in heated oil, or by vacuum techniques. Conventional P/M bearings can absorb from 10 to 30% by volume of oil. Solid resin impregnants are used to provide lubrication or to seal the pores, either to develop pressure tightness or to permit the use of processing solutions that would otherwise be trapped in the pores and degrade the component. Impregnation is also used in some cases to improve machinability.

4.1.3.3 Heat Treating Steel Components

The proper heat treatment of P/M components is an important concern, which should be considered at the design phase of the component. It not only affects the mechanical properties, but also imparts a dimensional change and a degree of distortion, which must be taken into account when designing tools.

Heat treatment is a generic term for a thermal cycle. There are many types of cycle that are designed to impart specific properties to the material through modification of the microstructure. Most involve heating into the austenitic phase region and cooling under various conditions to achieve a desired microstructure.

Heat treating procedures for P/M components have thermal cycles similar to those of wrought steels of similar composition. However, the response to the cycles are significantly different, primarily due to microporosity. Those processes commonly used are:

- Normalizing
- Annealing
- Stress relieving
- Neutral hardening
- Case hardening
    Carburizing
    Carbonitriding
    Nitrocarburizing
    Nitriding
- Austempering
- Quenching
- Tempering
- Gas nitriding
- Local hardening
    Induction
    Flame
    Laser

Normalizing and annealing provide a soft, ductile condition. Slow cooling rates are used to attain a ferritic/pearlitic microstructure, which is best suited for optimum machinability, welding and cold working.

Neutral hardening and case hardening utilize a rapid cooling or quench in oil from the solution temperature, converting the microstructure to martensite. The cycles impart the maximum strength and hardness. In wrought steel, these properties are governed by composition and grain size. In P/M, properties are also affected by density (percent microporosity).

Neutral hardening is commonly applied to Class 1 and Class 2 components where a press and sinter operation is sufficient to meet dimensional tolerance requirements. The steel is heated to its solution temperature in an atmosphere of controlled carbon potential to prevent both carburization and decarburization. The solution temperature is a function of the carbon content of the steel. Rapid cooling of the steel by immersion in a quenchant fluid causes transformation of the austenite, which is present at the solution temperature, into martensite. Higher carbon contents provide hardenability to form martensite on quenching. Low carbon content is normally specified when the application requires coining, sizing or machining. Other types of neutral hardening are induction and flame, which are rapid heating cycles usually applied with specialized equipment to individual components rather than a batch lot.

Carburizing and carbonitriding are the two most commonly used case hardening heat treatments. Carburizing utilizes a furnace atmosphere with a higher carbon potential than the base steel composition. A carbonitriding atmosphere contains high levels of both carbon and nitrogen, which impart maximum hardenability, and compressive residual stresses, which enhance fatigue strength. These treatments are normally specified when maximum fatigue strength and impact energy are desired. Density should be above 7.0 g/cm^3 (89% relative density) to attain the best combination of properties.

In nitrocarburizing and nitriding, the major component of the atmosphere is nitrogen. The treatments are performed at temperatures below 720°C (1330°F), and rely on diffusion of nitrogen to achieve surface hardening, rather than transformation to martensite. They are classified as surface treatments since no changes occur to the core microstructure of the component. They are normally applied where resistance to sliding wear is the primary requirement.

4.1.3.4 Machining

P/M is normally used to minimize machining operations by incorporating as much of the required finished configuration as possible. Machining is sometimes necessary to produce special shapes, threads, cross holes and closer tolerances. Machining parameters for P/M components are different from castings. Materials can be added readily to the powder blend to improve machinability. Small additions of sulfur and manganese sulfide are common in ferrous components, and lead is occasionally added to nonferrous components. Copper infiltration of steel parts also improves machinability. Oil or resin impregnation improves machinability of all porous P/M components.

Machining of high density P/M parts (above 92% dense) is similar to machining wrought metals. Lower densities require adjustments for optimum results.

Machining procedures for self lubricating porous components should avoid closing surface microporosity. Sharp tools and light cuts should be used in single point machining. While cutting fluids are preferred for most machining operations, fluid pickup increases with greater component microporosity. Fluids should be chemically compatible with the P/M alloy; subsequent processing may make it necessary to remove them. Ideally all machining, except grinding, should be done before deburring because retained deburring abrasives can cause excessive tool wear.

Following are general guidelines for performing various machining operations, when required.

1. **Turning.** Sharp pointed tools are recommended. Carbide is used for softer materials and titanium nitride coated carbide or cubic boron nitride for harder materials. Cutting angles on single point tools should be 15°, back rake about 10°, and reliefs about 10°. Cutting speeds of 0.8 to 1.6 m/s (175 to 350 sfpm) and feeds of 0.05 to 0.10 mm (0.002 to 0.004 in.) are normally used. Depth of cut is preferably at least 0.10 mm (0.004 in.) to avoid surface rubbing, which work hardens the material and accelerates tool wear. When closing of surface microporosity must be avoided, slow speeds of 0.5 to 0.75 m/s (100 to 150 sfpm), lower depth of cut of 0.05 mm (0.002 in.), and very sharp tools are used.

2. **Milling.** Cutters of high speed steel and carbide are recommended. Helical cutters with axial rake are preferred so that chips are sheared on an angle. Recommended speeds are 0.36 m/s (70 sfpm) for high speed steels and up to 1.52 m/s (300 sfpm) for carbide tools. Feeds should be 0.05 to 0.13 mm (0.002 to 0.005 in.) per tooth for roughing and 0.03 to 0.05 mm (0.001 to 0.002 in.) per tooth for finishing.

3. **Drilling and Tapping.** Speeds and feeds used for drilling P/M components are usually 80 to 85% those for wrought metals of the same composition. Nitrided steel, high speed steels and carbide drills are recommended for long life. When holes in the direction of pressing must be drilled, they can be spotted, partially formed or chamfered in pressing the component, thus simplifying the drilling operation. Internal threads are produced by conventional tapping methods. Premium grade steel and nitrided taps give excellent life. Titanium nitride coated taps may be used on harder materials. Roll form tapping is also used, especially in blind holes. P/M design can offer a process advantage by providing a burr relief slot (Figure 4-13).

4. **Grinding.** Grinding operations similar to those for wrought products apply to P/M components. However, where surface microporosity is required, grinding may be unacceptable because it tends to reduce or close microporosity. For hardened materials and high precision, cubic boron nitride grinding wheels are very effective.

4.1.3.5 Surface Treatment

Practically all common finishing methods apply to P/M components. While procedures in many applications are similar to those for wrought and cast components, the structural characteristics and microporosity, particularly in low density components, require some modifications in finishing operations. Following are descriptions of various surface treatments applied to P/M components.

1. **Deburring.** P/M components are rolled in rotating barrels or agitated in vibrating tubs (usually with some form of abrasive media and water) for cleaning, deburring, polishing, rolling or burnishing. Rust inhibitors should be added to the water. Components may be spun dry after the operation; in addition, heat may be used to evaporate water from the pores.

    Components can be resin or oil impregnated before tumbling to eliminate possible water absorption during deburring. Deburring should be done after machining to avoid abrasive pickup in the pores that could cause excessive tool wear. Residual abrasives should be avoided on components where carryover could be detrimental to the functioning of P/M components in service.

2. **Burnishing.** Burnishing can improve component finish and dimensional accuracy or work harden surfaces. Tumbling, roller burnishing, ball sizing and stick burnishing, using a broach-like tool with buttons or ridges, are common burnishing techniques. As with machining, precautions may be required to avoid closing desired surface microporosity.

3. **Blackening (bluing).** Ferrous P/M components can be colored by several methods. They can be furnace blackened to impart indoor corrosion resistance. Oil dipping gives a deeper color as well as slightly greater corrosion resistance. An oil that leaves a dry film on the components, may be used.

   Ferrous P/M components can also be chemically blackened, using a commercial liquid salt bath. Entrapment of salt in the pores of components should be avoided. The components can be impregnated, prior to blackening, with a resin that will not break down in the bath. Alloys containing nickel and copper tend to adversely affect most blackening baths. These materials also seriously affect component color. An oil dip improves appearance and corrosion resistance as it does with furnace blackening.

4. **Steam Treating.** Steam treating forms an oxide coating (Fe_3O_4). It is commonly used to improve the wear properties of ferrous P/M components and additionally provides improved corrosion resistance. The process closes some of the interconnecting microporosity and all surface microporosity, increasing component density and compressive strength, and providing gas pressure tightness 1700 MPa (250,000 psi). The process tends to cause a slight size change (0.0075 to 0.0125 mm or 0.0003 to 0.0005 in. pickup per surface) and it makes the components somewhat less ductile and more difficult to machine.

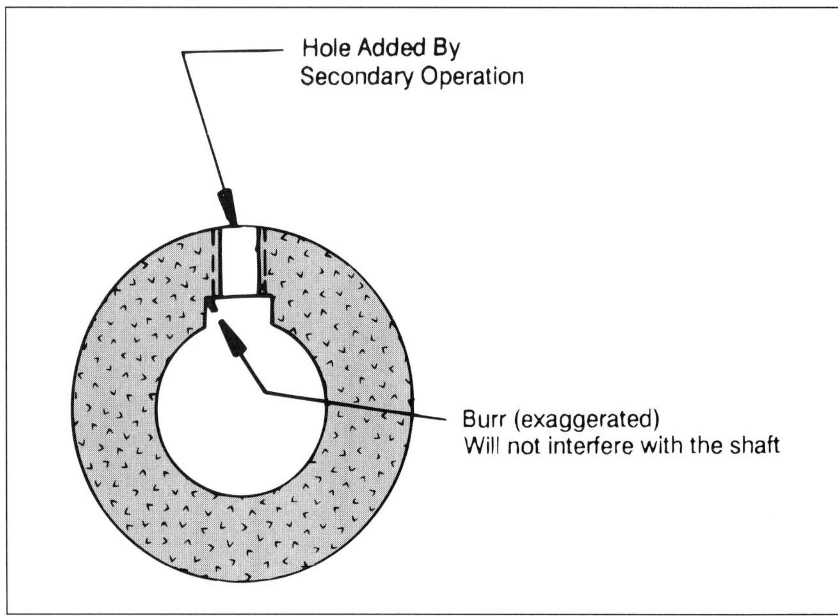

Figure 4-13 Burr relief slot.

Components are heated to 480 to 595°C (900 to 1100°F) and exposed to superheated steam under pressure. They can be oil dipped after cooling to increase corrosion and wear resistance further and enhance appearance. Heat treated components are not steam treated since the treatment will alter the properties achieved in heat treatment.

5. **Plating.** All types of plating in general use, including copper, nickel, chromium, cadmium and zinc, can be applied to P/M components. Components should have microporosity sealed, usually by resin impregnation, to avoid entrapment of plating solutions in the pores. Components that have been oil impregnated or oil quenched must have all oil removed from the pores and surface prior to resin impregnation and/or plating. Electroless nickel plating also can be used, and peen (mechanical) plating is applicable to non-impregnated components.

4.1.3.6 Welding/Brazing

Ferrous P/M components can be successfully welded or brazed if required. Most conventional welding methods (TIG, MIG, electron beam, resistance, projection and friction) are applicable. Welding performance is chiefly determined by component density, alloy composition and, most important, proper design.

Microporosity, or relative density, is most influential in fusion welding. Higher density components can generally be joined using welding parameters similar to those selected for wrought or cast counterparts. However, welding lower density components often results in a greater degree of shrinkage in the weld zone because of particle melting. The subsequent solidification of weld metal causes local stresses that often initiate cracks. Microporosity can also cause erratic fluctuations in welding performance if non-metallic materials become entrapped in the voids. Proper design and process procedures should eliminate the following potential contaminants, which may reduce weld integrity.

- Residual lubricants
- Quench oils
- Machining coolants
- Plating solutions
- Impregnating materials
- Cleaning or tumbling agents
- Free graphite or residual ash

Many P/M compositions can be welded without difficulty. However, some elemental additions or grades should be avoided where possible. Carbon content has a pronounced influence on overall weldability. As a general rule, the carbon content should be held to as low a level as possible consistent with strength requirements.

Materials containing elemental sulphur additions should be avoided, as it can migrate to the grain boundaries and may cause cracking during fusion welding. If a machining enhancement is necessary, an addition including manganese sulphide (MnS) would be more appropriate.

Parts made from premixes with copper additions of two per cent or less can be joined readily by most welding processes. Compositions that include both sulphur and copper are exceptions. Admixes of nickel to iron or low alloy steel powders generally enhance toughness and do not pose any welding difficulties.

Stainless steel P/M components have been successfully welded using various processes. The 303 free machining grade and those identified as nitrogen strengthened are not good candidates for fusion welding applications. The 410 martensitic grade can be welded, but precautionary measures must be observed with regard to the hardenability of the material.

To ensure a successful P/M weldment, three factors must be considered: type of joint, material selection and joining process. The primary factor in joint design is to ensure that the weld interface is located in an area that is not subject to high loads or stress concentrations.

The material must satisfy the strength requirement of the application. If this consideration necessitates materials that exhibit high hardenability, the joining process must permit control of the hardness of the weldment and prevent cracking.

The joining process must be considered when evaluating the strength requirements of the application. The fusion processes, typically MIG, TIG and laser beam welding, provide the highest joint strengths. Friction and resistance projection welding are often nearly equal. Brazing, diffusion and adhesive bonding provide somewhat lower joint strength.

P/M components are often joined by furnace brazing. This method allows P/M fabricators to use existing sintering furnaces and atmospheres to join P/M components to one another or to wrought or cast materials.

When choosing a brazing alloy, capillarity of the P/M pores imposes special considerations. Standard copper brazing materials wick into the pores at the surface, leaving insufficient filler to establish a satisfactory bond. Brazing systems have been designed for P/M that restrict brazing alloy penetration to the immediate adjoining P/M surface areas.

The preferred joint for brazing is a lap rather than a butt. Thermal expansion of the materials being joined must be considered in order to maintain the proper gap distance at temperature. Ring preforms placed adjacent to the joint or the spot location of a brazing paste are preferred to sandwich positioning of the filler between components. Fixturing may be necessary for more complex configurations or when tighter dimensional controls are required. Brazed components can be subsequently quenched and tempered or steam treated.

4.2 METAL INJECTION MOLDING (MIM)

The MIM process is used to produce components with properties equal to or better than those achieved by P/M, equivalent to those achieved by powder forging, and with much greater complexity than is possible with either. The process is performed in four steps.

1. <u>Feedstock preparation</u> Fine metal powders are intimately mixed into a thermoplastic carrier at an elevated temperature, hardened by cooling to room temperature and granulated to produce a feedstock.
2. <u>Injection molding</u> The feedstock is heated in the feed barrel of an injection molding machine (consistency of toothpaste) and injected into a mold to form a green body of the desired shape. The green body is cooled to harden and ejected from the mold.

TABLE 4-3 CHARACTERISTICS OF MIM VERSUS P/M

	P/M[a]	MIM
Particle Size	20-250 μm[b]	< 20 μm
Particle Response During Shaping	Plastic Deformation	Undeformed
Porosity Before Sintering	10 - 20%	30 - 40%
Amount of Binder/Lubricant	0.5-2%	30 - 40%
Homogeneity of the Green Body	non homogeneous	homogeneous
Final Sintered Density	$\leq$ 92%	> 96%

a Typical Values
b A small percentage may be below this value

3. <u>Debinding</u> A major portion of the binder is removed without disturbing its shape.
4. <u>Sintering</u> The porous green body is raised to an intermediate temperature in a sintering furnace to remove the remainder of the binder, then sintered at high temperature to develop the anticipated mechanical and physical properties.

The characteristics of MIM compared with P/M are summarized in Table 4-3.

4.2.1 Feedstock Preparation

Feedstock preparation is critical to the final dimensional accuracy of the MIM component. Pore volume is occupied by the thermoplastic binder, which is incompressible. Therefore the green density, and consequently shrinkage during sintering, is determined by a controlled mixing process, in which the ratios of thermoplastic carrier and metal powder are held within very close limits. The feedstock is normally in the range of half metal powder and half binder by volume. Powder size is very fine, as noted in Table 4-3 above.

The feedstock is 100% recyclable in-house. In normal molding practice the sprues, runners and gates, which are trimmed off of the ejected mass, are recycled in-house. Rejected green bodies are likewise recycled.

4.2.2 Injection Molding

Injection molding is performed on essentially the same machines and tooling that are employed for injection molding of plastics and composites. During injection, typically at low pressures of 70 MPa (10,000 psi), the feedstock flows around corners, into undercuts and into thin sections. The molds employ side cores, rotating tool members and other creative concepts to produce very complex shapes, similar to those produced by plastic injection molding, and much more complex than by P/M or powder forging.

The metal content of the feedstock affects molding parameters such as thermal diffusivity (thermal conductivity divided by heat capacity and density) and flow characteristics. Molding parameters are adjusted accordingly. Generally, mold temperatures are higher, feedstock temperatures lower, injection pressures lower and injection speeds are much lower than for conventional plastics. Ejection pins, runners and gates are larger.

The design of the mold must account for shrinkage, which is typically 20% in all directions. Shrinkage varies somewhat from the theoretical, which is predicted from microporosity, depending on the flow of feedstock through the mold cavity. Experienced mold makers can predict variations closely. Where very close precision is required, mold makers may modify the molds after the initial trial run. All design details are retained after shrinkage, and tolerances of approximately ±0.3 - 0.5% of the total dimension are typical.

4.2.3 Debinding

There are two methods of debinding in use. One is a two-stage process that begins with solvent extraction, which removes some of the binder. The remainder of the binder is burned out in a debinding furnace. The other method is evaporative. It requires a binder that exhibits the required stability and flow characteristics for molding, but can be easily evaporated during debinding.

Both processes must be carefully controlled so as to preserve product integrity, avoiding potential problems such as slumping, cracking, spalling and fracturing. The very narrow process limits dictate a relatively long debinding time under close control. In both processes, some of the binder is deliberately retained to provide enough green strength for further handling. The debinding process also employs control of the atmosphere to minimize oxidation of the metal in the green body.

4.2.4 Sintering

The final step is a two-phase sintering operation in a programmable furnace with controlled atmosphere. Precise control using microprocessors is employed to assure product uniformity. In the first phase, the remaining binder is driven off and oxygen is removed. In the second stage, the temperature is elevated and sintering is accomplished. Carbon content of the component is controlled by controlling sintering time, temperature and atmosphere. The two-phase operation requires substantially more time than P/M sintering, which contributes to the increased cost of MIM compared with P/M.

Sintering produces components with relative densities greater than 96%, which is generally higher than with P/M. The microporosity is not interconnected. The high density contributes to improved mechanical properties, such as higher tensile strength, and substantially higher ductility and impact energy.

4.2.5 Secondary Operations

The capability of MIM to produce very complex shapes and close tolerances reduces, and often eliminates, finish machining operations. Where required, a variety of secondary operations can be performed at all production stages. Processing can be performed in the green stage, but operations such as machining must be done with care because the green bodies are very fragile. If green bodies have similar processing parameters, they may be welded by placing the desired surfaces together during debinding.

After sintering, machining operations such as broaching, drilling, tapping, turning, grinding, lapping and burnishing are performed as readily as with equivalent cast or wrought alloys. Residual microporosity is very low, uniformly fine, and evenly distributed with no large pores, and interconnected microporosity is usually less than 0.2%. Therefore, conventional surface treatment processes can be employed without impregnating. Components may be welded or heat treated similar to equivalent cast or wrought alloys. Appropriate alloys can be case hardened to closely controlled depths. Because of the very low level of interconnected microporosity, MIM components are not impregnated with resin or oil or infiltrated with copper as is often done with P/M.

4.2.6 Trends in MIM Processing

Ongoing improvements in all stages of the MIM process are reducing costs, increasing the range of alloys that can be used and permitting increased component size. Debinding times are being sharply reduced, producing a reduction in both processing cost and oxidation. Oxidation is being further reduced with new technology in the control of the debinding and sintering atmospheres. MIM producers are also developing sintering process that utilize coarser powders and still reach the high densities that formerly required very fine powders. The net effect is a continuous reduction of MIM processing cost, a wider range of available alloys, and potentially larger components.

4.3 POWDER FORGING (P/F)

The P/F process is used to produce components with nearly full density, and properties equivalent to those developed by conventional precision forgings made from billets. The process is performed in three steps.

1. <u>Preform compaction</u> Metal powders are compacted in a process similar to conventional P/M compaction to produce a preform with very close control of mass and density.
2. <u>Preform sintering</u> The preform is sintered in a process similar to P/M.
3. <u>Forging</u> The preform is forged to final density and/or form, either by hot repressing or hot upset forging.

4.3.1 Preform Compaction

The compaction process is similar in principle to conventional P/M compaction. However there are several differences, which are critical to the forging operation. If the preform is to be finished by hot repressing, it will have essentially the same shape as the finished P/F component. If features are to be formed by hot upset forging, the preform will have a simpler shape. In either event, the mass and density of the preform must be controlled to ensure a successful forging operation. Total powder mass is normally held to plus or minus one-half per cent to avoid tool overload. Density distribution is carefully controlled to achieve forging die fill with very high density, avoid forging defects and achieve the required dimensional precision.

In order for a component to be a candidate for P/F, its shape must allow for a preform in which density can be controlled. It is therefore essential for the product designer to work very closely with the P/F parts producer beginning with the preliminary design stage of product development.

Each application requires its own preform development, which is performed by the P/F producer. The preform shape is carefully controlled to develop shear strains that collapse the porosity with the minimum applied load. The correct amount of strain will ensure that all areas of the cavity fill properly, develop a preferred grain flow in the forging, and collapse porosity to the maximum amount feasible. Excessive deformation may

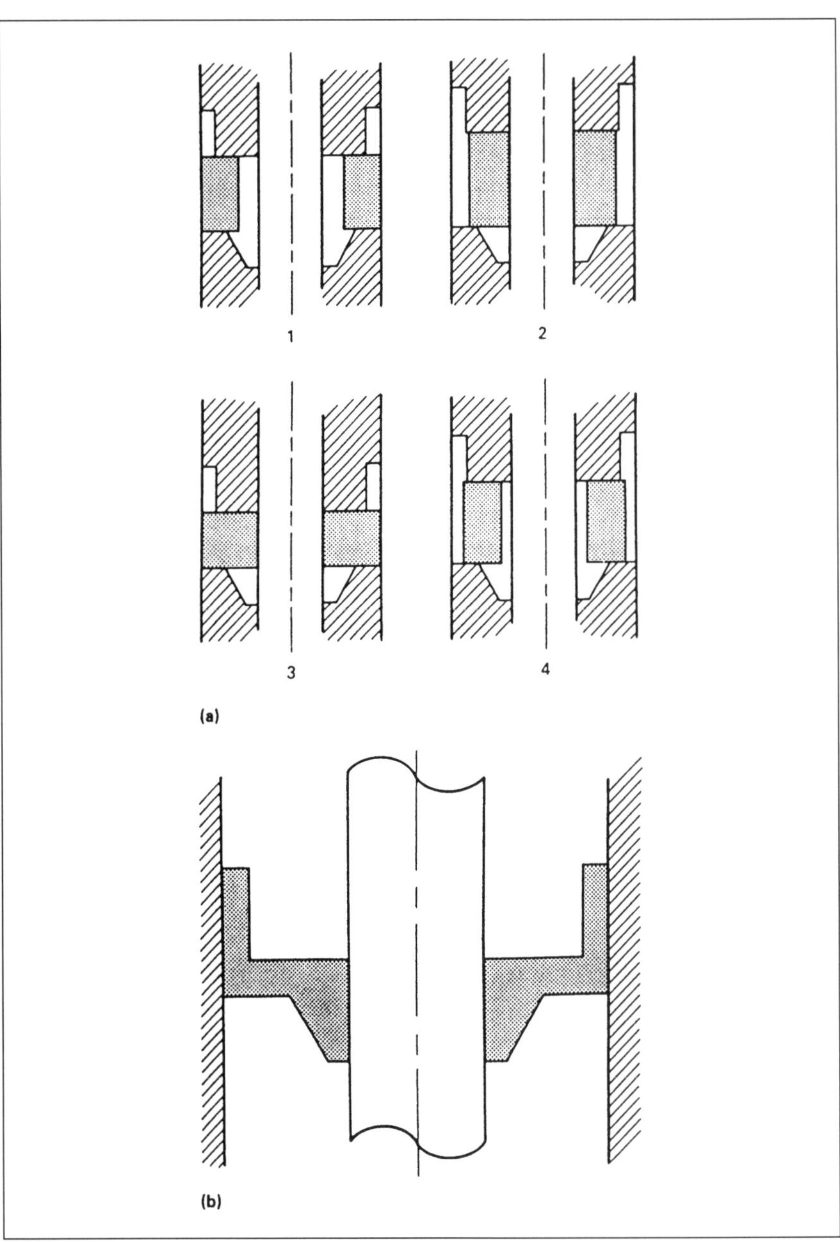

Figure 4-14 (a) Possible configuration for the ring preform for forging the part shown in (b). See text for details. (b) Cross section of the part under consideration for powder forging.

exceed the fracture limits of the porous preform, and develop internal flaws.

Successful preform designs have been based on experience, and refined by iterative trials. Computer aided design (CAD) is now being used to facilitate preform design. For example, Figure 4-14 illustrates four different preforms that were candidates to make the axisymmetric component with the hub and cup features. Each choice utilizes a different member in the forging tools to position the preform, and each produces a different amount and combination of forward extrusion, backward extrusion and upsetting. The use of preforms 2 and 4 resulted in cracks at the outer rim due to undesirable metal flow around the corner of the leading surface on the upper punch. Preform 3 was found to be prone to produce cracking at the top surface of the hub. Preform 1 was successfully used to produce defect-free forgings.

4.3.2 Preform Sintering

Sintering requirements for forging are similar to those for P/M with one exception: the emphasis on chemical refinement is much greater. Control is essential to minimize residual oxides, inclusions and nonuniform alloy distribution, which can act as stress raisers.

4.3.3 Forging

Forging densifies the preform, reducing porosity to very low levels and improving mechanical properties. There are two types of forging operations, which are illustrated in Figure 4-15: hot repressing and hot upset forging. In the hot repressing process, the preform is essentially the shape of the finished forging. Repressing densifies the preform, thereby improving mechanical properties, bringing dimensions to their final tolerance and developing the surface finish.

Hot upset forging causes a calculated amount of plastic deformation. For example, teeth on a gear or sprocket may be formed from an annular preform by extruding metal into the tooth cavities as illustrated in Figure 4-16. The amount of plastic flow is largely governed by the shape of the preform.

There are two differences between P/F forging and conventional forging from wrought billets. First, the P/F preform is forged in a single press stroke, compared with the multiple hits, sometimes combined with multiple rolling operations, that characterize conventional forging. Second, plastic deformation in a P/F operation is much less than in conventional forging where all features are usually formed by extensive plastic deformation that is controlled by developing a relatively large amount of flash, which is subsequently trimmed off.

There are two types of P/F forging dies: open and closed. The open die, shown in Figure 4-17 is similar to an impression die used in conventional hot forging. It consists of upper and lower die halves, with impressions that form the die cavity. The dies form a flash gutter at the parting plane that receives metal from the cavity (flash), which is subsequently trimmed off and recycled. The closed die, shown in Figure 4-18 consists of upper punch, lower punch and die, which form the cavity and completely enclose the forging. There is no flash gutter, so that the forging is essentially flash free. Under some conditions, a thin flash may form in the clearance between the upper punch and the cavity, but this flash is easily removed.

The plastic flow achieved with hot upset forging develops better mechanical properties than repressing. However, it requires more preform

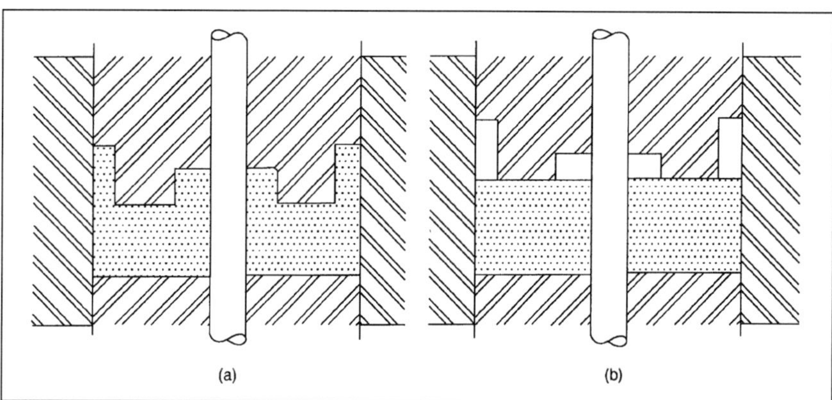

Figure 4-15 Comparison of (a) powder repressing and (b) powder forging.

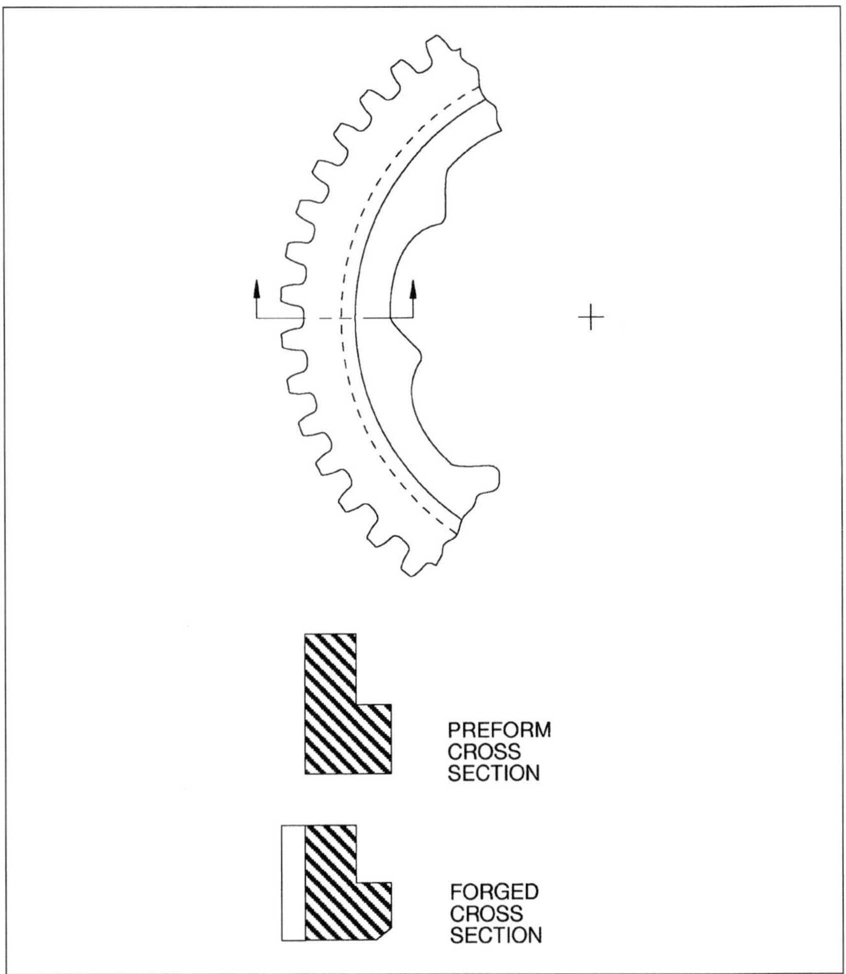

Figure 4-16 The external teeth and internal cams on this ring were formed in the forging operation from annular features in the blank. See also Figures 2.3-36 through 2.3-39 in Section 2.3.

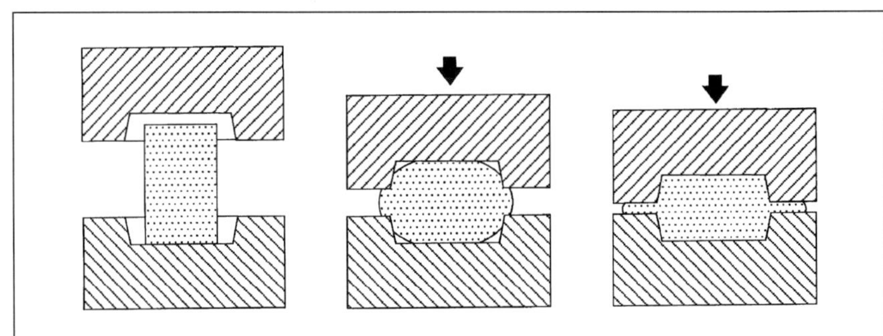

Figure 4-17 Impression die with flash for conventional forging

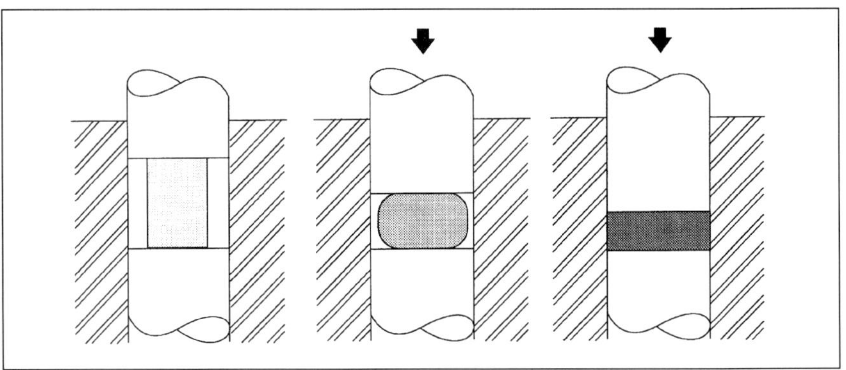

Figure 4-18 P/F trap die.

development and may incur higher die wear, particularly in areas of highest plastic flow.

4.3.4 Effects of Process on Material Properties

The material properties of a P/F component are process driven and, in some cases, process sensitive. Factors such as microstructure and chemistry are affected by process variables. For example, grain size is finer than for cast or wrought products, which affects the response to heat treatment. For steels, the fine grain size yields a small but distinguishable difference in hardenability. The improved uniformity of grain size produces a more uniform response to heat treatment.

Processing variables must be carefully controlled to minimize inclusions, oxides and microporosity, which reduce mechanical properties. The effects of these factors on fracture resistance, such as impact energy, fatigue strength and fracture toughness, have been documented. Inclusions are controlled by proper precautions in powder production, in sintering and in handling of the heated preform. Oxides are minimized in the sintering process, primarily through control of temperature and atmosphere.

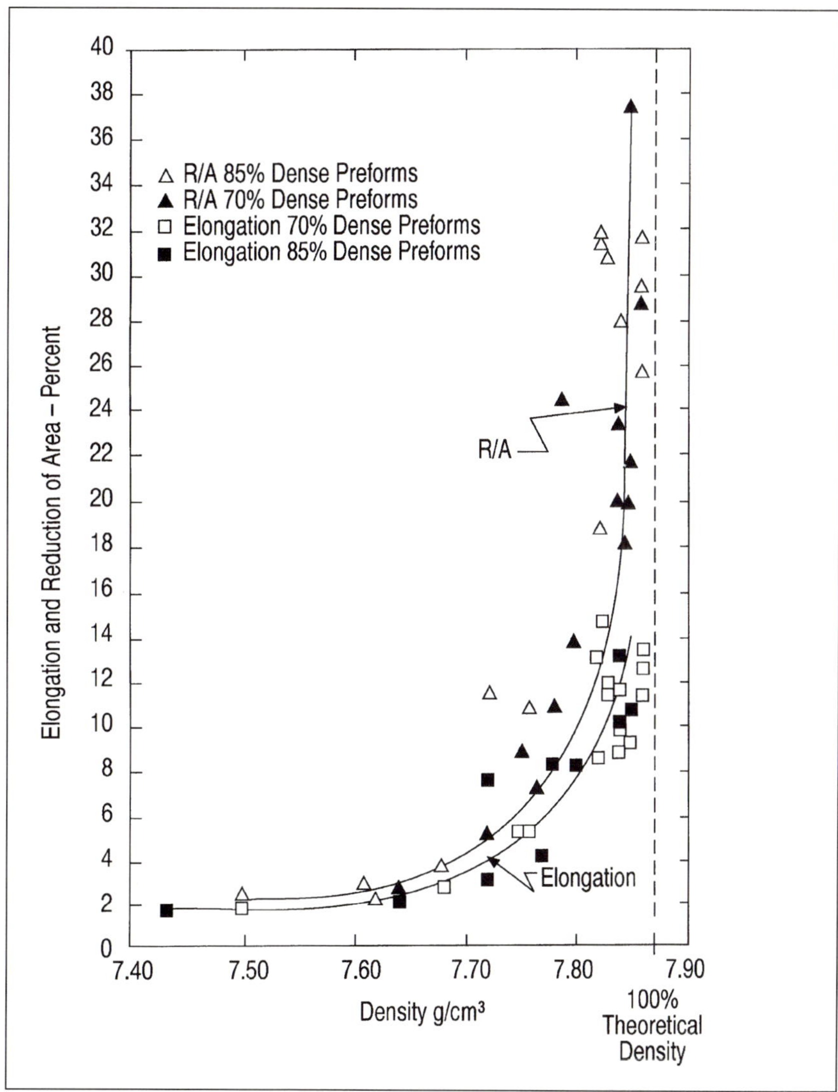

Figure 4-19 Effects of density on elongation and reduction of area.

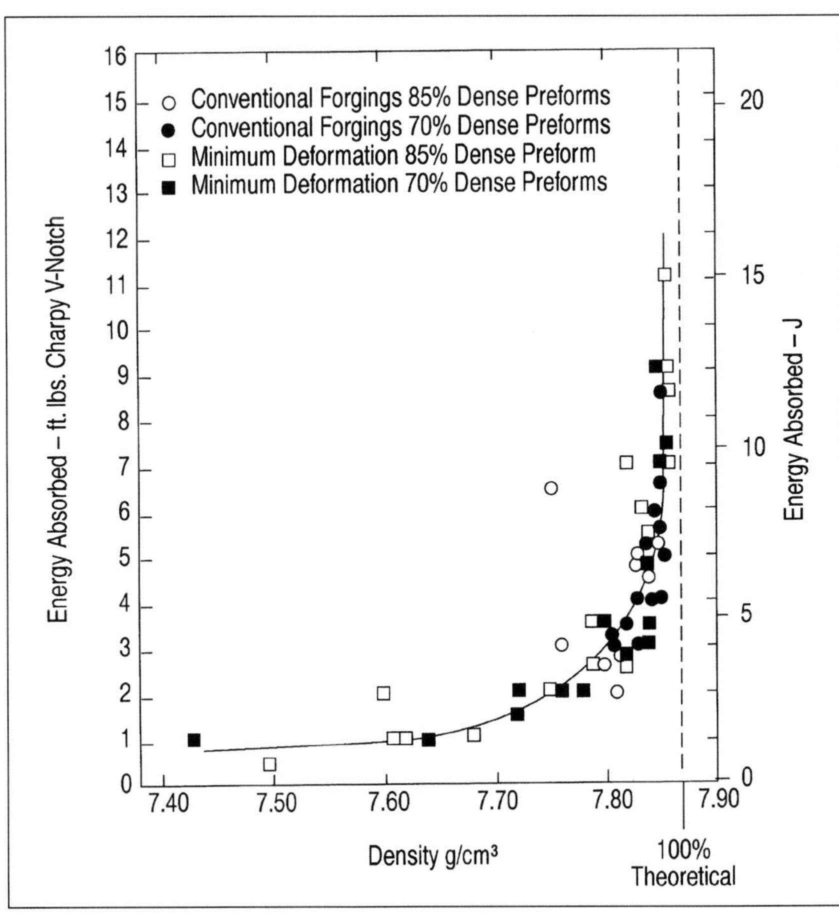

Figure 4-20 Effect of density on energy absorption.

Full density in the critical sections is the goal of powder forging, but residual microporosity is inevitable. Users frequently specify 99.5 per cent relative density as a minimum overall density level. P/F can achieve this level, but local levels may be higher in microporosity, while others may be virtually pore-free. Ductility, and consequently impact energy, are affected by microporosity, even at low levels. At very low levels, as the relative density approaches 100%, ductility and impact energy increase sharply as illustrated in Figures 4-19 and 4-20. In some applications, it may be necessary to specify maximum levels of microporosity in critical areas of a P/F component.

4.3.5 Secondary Operations

The very low level of microporosity in P/F components, which is not interconnected, allows operations such as finish machining, surface treatment, heat treatment and welding to be performed using processes that are similar to those of cast and wrought products of similar composition. Surface treatment processes do not require impregnation, as is done with P/M.

The response of P/F components to heat treatment is uniform due to the homogeneity of grain size and composition. The hardenability of P/F steels is somewhat less than that of conventional steels. Elements such as chromium and manganese form oxides during sintering that are difficult to reduce, so their use is limited in traditional P/F-42XX and P/F 46XX grades of steel powders. Higher molybdenum levels have improved the baseline hardenability of P/F grades to nearly that of AISI grades.

Case Studies

The preceding guidelines for the design of P/M components necessarily present conservative data and procedures. Adherence to the guidelines virtually assures that the anticipated performance can be consistently met in components produced by a qualified P/M parts producer.

The following section of case studies illustrate how the design guidelines have been interpreted in real world applications, and the contribution of the P/M parts producer. Some case studies additionally illustrate ways in which the limitations of the guidelines have been exceeded by the special expertise of the P/M parts producer to gain a product advantage.

No.	Title	Process
1	Crankshaft Sprocket for Automobile Engine	P/M
2	Air Bag Sensor Insert Assembly	MIM
3	Connecting Rod for Automobile Engine	P/F
4	End Plates for Automotive Power Steering Pump	P/M
5	Transmission Output Shaft Hub	P/M
6	Differential Shifter Sleeve	P/F
7	Throttle Lever Assembly	P/M
8	45 Degree Vari-Drive Cam Assembly	P/M
9	Copper Heat Sink	P/M
10	Self-Adjusting Mechanically Applied Brake Actuator System	P/M
11	Electrical Connector for Surgical Instrument	MIM
12	Crimping and Cutting Tool Components	MIM
13	Drive Hub for Office Machines	P/M
14	Brush Box for Permanent Magnet Motors	P/M
15	Electrical Contact Pivot	P/M
16	High Strength Brass Yoke for Sprinkler System	P/M

CASE STUDY NO. 1

Crankshaft Sprocket for Automobile Engine
Process: P/M

Component name:	Crankshaft Sprocket
Size - mm (in.):	61 (2.4) pitch diameter
Weight:	0.170 kg (0.37 lb)
Alloy:	Nickel Molybdenum Copper Steel
Tensile strength:	1170 MPa (170,000 psi)
Apparent hardness:	35 HRC
Particle hardness:	55 HRC (converted from Vickers HV0.5)
Density, min:	
g/cm^3 (lb/in.^{3})	6.8 (0.245) min, 7.0 (.250) at teeth
Secondary Operations:	Grind hub inside diameter
	Tumble deburr
Heat treatment:	Sinter furnace hardened and then tempered
Alternate process:	Forge, machine and induction harden
Annual Production:	800,000

The crankshaft sprocket, which drives the timing chain in a high-volume passenger car engine, requires a combination of high strength, resistance to wear and dimensional precision at minimum cost.

Strength is verified by a static load in which a load of 4900 N (1100 lb) is applied to the teeth at the operating contact point. Wear resistance is required to drive the timing chain with consistent precision and freedom from noise and vibration over the life of the engine. Both strength and wear resistance are achieved by a nickel molybdenum copper steel alloy hardened in a furnace sintering operation to develop a tensile strength of 1170 MPa (170,000 psi) and a particle hardness of 55 HRC. Hardening is accomplished in the sintering operation.

Dimensional precision is required to attach the sprocket to the crankshaft and drive the chain with minimal noise and vibration. The hub bore is precision ground to a total diametral tolerance of 0.025 mm (0.001 in.) and press fitted to the engine crankshaft. All other tolerances are held as compacted and sintered, with no finish machining operations. Torque is transmitted by a woodruff key, which engages a keyway in the hub. Keyway width is held to a total tolerance of 0.050 mm (0.002 in.) and an axial straightness tolerance of 0.15 mm (0.006 in.). The teeth are concentric with the hub bore to within 0.050 mm (0.002 in.) total runout.

The sinter furnace hardened sprocket displayed better dimensional accuracy and wear resistance than a P/M sprocket with conventional induction hardening. It offered a substantial cost reduction compared with the alternate process of forging, machining to dimensional tolerances and induction hardening.

CASE STUDY NO. 2

Air Bag Sensor Insert Assembly
Process: MIM

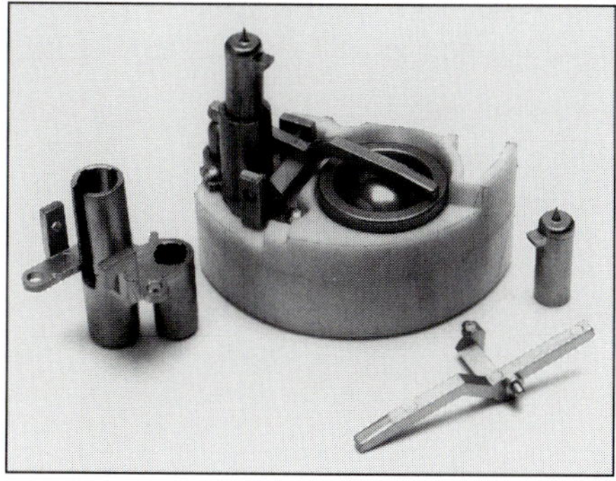

Component name:	Air bag sensor insert assembly
Alloy:	17-4 PH stainless steel
MPIF specification:	MIM 17-4 PH
Tensile strength:	1170 MPa (170,000 psi)
Yield strength:	1100 MPa (160,000 psi)
Hardness:	38-41 HRC
Elongation:	7.0%
Density:	
g/cm^3 ($lb/in.^3$)	7.60-7.68 (0.275-0.277)
Relative density:	97.7-98.8% (based on 7.78)
Secondary operations:	Heat treatment
Number of parts:	3 per assembly
Alternate process:	Investment casting

The insert assembly consists of three parts: D-shaft, tab insert and firing pin. It is a integral feature in an all-mechanical automotive air bag sensor. The material and process were chosen to achieve dimensional precision, rigidity, strength, wear and corrosion resistance. The end user and MIM producer worked together to achieve a "design for technology".

For the D-shaft, perpendicularity of the journal axis to the 33.6 mm (1.32 in.) long D-shaft arms is held to 0.050 mm (0.002 in.). The journal diameter of 1.45/1.49 mm (0.057/0.059 in.) with a surface finish of 0.2 micrometers is critical to proper function. For the firing pin, the diameter is held to 5.00/4.95 mm (0.198/0.195 in.) and the length to 15.0 ± 0.2 mm (0.59 ± 0.008 in.). For the tab insert, the center line of the bearing axis to the centerline of the barrel diameter is 3.88 ± 0.10 mm (0.153 ± 0.004 in.).

Metal injection molding provided a better surface finish and a higher tensile strength than investment casting. It also offered the capability to combine parts, thereby reducing the number of parts, eliminating fasteners and reducing assembly operations.

CASE STUDY NO. 3

Connecting Rod for Automobile Engine
Process: P/F

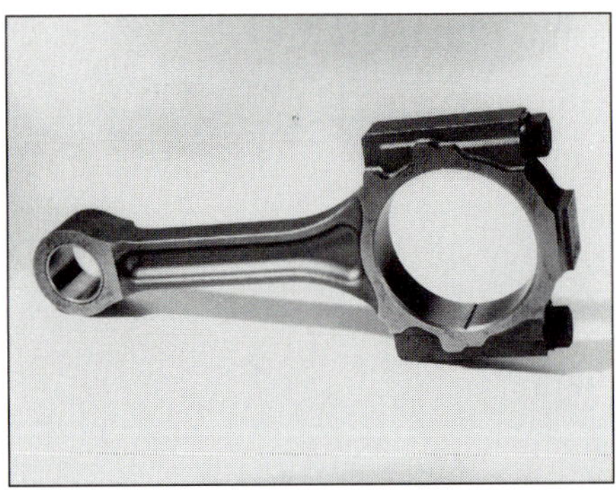

Component name:	Connecting Rod
Size - mm (in.):	210 (8.2) overall length
Weight - kg (lb):	0.75 (1.65) as forged
	0.66 (1.45) finished
Alloy:	Copper-Steel (MPIF FC-0205)
Tensile strength, min:	760 MPa (110,000 psi)
Yield strength:	550 MPa (80,000 psi)
Hardness:	24 HRC
Density:	
g/cm³ (lb/in.³)	7.84 (0.283)
Relative density:	99.6%
Secondary operations:	Deflashing, shot peening and finish machining
Alternate process:	Precision forging
Annual Production:	More than 2,000,000

The P/F connecting rod is used on a new generation of high performance V-8 passenger car engines. P/F produces rods with mechanical properties equivalent to those of conventional precision forging. For example, the mean endurance limit is 275 MPa (40,000 psi) with a standard deviation of less than 5%. This value is 153% of the stress level when the engine is operating at the maximum functional design loads. The high level of fatigue performance is achieved by forging to 99.6% of relative density, virtually eliminating microporosity, and by maintaining a high level of powder purity.

P/F also produces rods with much closer tolerances than their precision forged counterparts. For example, dimensional variation on the outside profile of the rod is held to 0.12 mm (0.005 in.) compared with 0.5 mm (0.020 in.) for conventional precision forged rods. There were similar advantages in straightness and in elimination of surface defects. The dimensional precision achieved by P/F eliminated a broaching operation on the bolt bosses. A minimal balancing pad is provided on the rod end, and no balancing pad is required on the pin end. The cap is fractured from the rod, eliminating saw cutting, face grinding and line reaming operations that are normally required. The fracturing operation leaves very fine mating irregularities in the rod and cap, which locate the members securely during reassembly and eliminate the tendency to shift in service.

The rods also display a uniform microstructure with a structural integrity that has eliminated the need for nondestructive test procedures such as magnetic particle inspection.

CASE STUDY NO. 4

End Plates for Automotive Power Steering Pump
Process: P/M

Component name:	Power steering pump end plates
Size - mm (in.):	54.5 (1.79) diameter
Weight, kg (lb):	0.115 (0.25) and 0.182 (0.40)
Alloy:	Carbon steel
MPIF specification:	F-0005-15
Tensile strength, min:	180 MPa (26,000 psi)
Yield strength:	150 MPa (22,000 psi)
Hardness, minimum:	70 HRB
Density, min:	
g/cm^3 (lb/in.^{3})	6.1 (0.22) min., 6.4 (0.23) typical
Relative density:	78% min, 81% typical
Secondary Operations:	Steam treatment, Resin impregnation
Machining operations:	Drill holes, turn O.D., grind wear face
Surface treatment:	Painting
Annual Production:	4,000,000

P/M proved to be the only feasible process for manufacturing the end plates, which are required to meet stringent performance requirements, including:

- Wear resistance
- 100% sealing, after machining
- Good machinability
- Static and fatigue strength to resist oil pressure peaks and subsequent high frequency cycling
- Close control of critical features

In order to achieve the required wear resistance, the end plates are carburized during sintering. This operation increases the carbon content on the wear surface from a range of 0.4 to 0.6% in the core to a range of 0.6 to 1.0% on the surface to a depth of 1.3 mm (0.033 in.). Wear resistance is further improved by a steam treating operation.

Sealing is required to prevent leakage of hydraulic fluid under high pressure. It is partially achieved by the steam treatment, and is completed by resin impregnation. Static and fatigue strength are verified by applying static proof loads of 10,000 N (2200 pounds).

Several features are critical to oil flow. They are maintained to the required intricacy and dimensional precision through careful tool preparation and by maintaining specific compactability properties of the powder.

CASE STUDY NO. 5

Transmission Output Shaft Hub
Process: P/M

Component name:	Transmission output shaft hub
Size - mm (in.):	130 (5.12) O.D.
Weight kg (lb):	0.645 (1.42)
Alloy:	Iron-nickel
MPIF specification:	FN-0205 (Modified)
Tensile strength:	
MPa (psi)	1240 (180,000)

Density, minimum:	Overall	Spline
(g/cm^3) $(lb/in.^3)$	7.1 (0.25)	7.3 (0.26)
Relative density, %	90	93

Secondary operations:	Turning in the soft state, hard turning and grinding
Heat treatment:	Batch carburizing and tempering
Alternate process:	Wrought steel
Annual production:	160,000

The output shaft hub is used in transmissions for rear-wheel drive, full size and luxury automobiles, where it transmits power directly to the output drive shaft. A radial crush load test specification of 18.4 kN (4150 lb) and an axial crush load test of 43.1 kN (9,700 lb) are imposed to ensure structural integrity. Because the walls between the internal spline major diameter and the boss diameter are thin, they were designed to minimize stress intensity factor.

Close control of density distribution and high dimensional precision are essential to the performance of the hub. The components are double pressed and double sintered to achieve high density. Tolerances are being held on the boss diameter to ±0.010 mm (0.0004 in.) and on the counter bore diameter to ±0.019 mm (0.0008 in.). Flatness is held on the two thrust faces to 0.025 mm (0.001 in.) with a surface finish of 0.375μm (15 microinches).

Wrought steel was considered as an alternative. However, wrought steel required substantially more finish machining operations than with P/M, which meant a large capital investment in a machining line plus the operating cost.

CASE STUDY NO. 6

Differential Shifter Sleeve
Process: P/F

Component name:	Differential shifter sleeve
Size - mm (in.):	80 (3.15) O.D.
Weight - kg (lb.):	0.454 (1.00)
Alloy:	Nickel-molybdenum steel
Tensile strength, min:	2070 MPa (300,000 psi)
Hardness, minimum:	
Surface	57 HRC
Core	55 HRC
Density, min:	
g/cm^3 (lb/in.^{3})	7.82 (0.283)
Heat treatment:	Induction through-hardened
Alternate process:	8625 steel conventional forging

The sleeve functions to engage the front axle when the vehicle is shifted from two-wheel to four-wheel drive. It was previously made from an 8625 conventional forging, which required extensive finish machining operations. The internal involute spline is hot forged to a size that accommodates material growth during hardening. Major and minor diameters are held to a tolerance of 0.25 mm (0.010 in.). The spline tolerance between the maximum actual and minimum effective circular space width is held to 0.18 mm (0.007 in.).

In its current form, the sleeve illustrates the capability of the powder forging process to produce splines as well as teeth and pointing to very close tolerances.

CASE STUDY NO. 7

Throttle Lever Assembly
Process: P/M

Component name:	Throttle lever assembly
Overall - mm (in.):	128 (5.0)
Weight g (oz):	105 (3.7)
Alloy:	Copper-steel
Strength - MPa (psi):	
Tensile	345 (50,000)
Yield	310 (45,000)
Density:	
g/cm³ (lb/in.³)	6.6 (0.24)
Number of parts:	2—Machined pin sinter brazed to P/M lever
Alternate process:	Assembly of blanked and formed and machined parts
Annual production:	25,000

The lever attaches to the throttle linkage on a diesel engine used in trucks, marine and industrial applications. Several critical dimensions are held to close tolerances. A bore diameter of 15.00 mm (0.5906 in.) is held to ±0.05 mm (0.002 in.), and a 22.90 mm (0.9016 in.) diameter boss is held to ±0.13 mm (0.005 in.). The mating diameters in the pin and lever are controlled to ±0.04 mm (0.0016 in.).

The assembly passes three tests. The first applies a transverse load of 900 N (200 lb) to the pin. The second applies a transverse test of 450 N (100 lb) to the 9.90 mm (0.390 in.) diameter boss. The third applies an axial load of 300 N (67 lb) to the pin to test the sinter brazed joint. The lever achieves the required hardness without requiring heat treatment.

The lever had been formerly made from a blanked and formed lever brazed to two machined bosses and a machined pin. The cost was reduced by 73% by going to the P/M assembly, and both quality and appearance were improved.

CASE STUDY NO. 8

45 Degree Vari-Drive Cam Assembly
Process: P/M

Component name:	Vari-drive cam assembly
Size - mm (in.):	125 (4.92) O.D.
Weight - kg (lb):	2.26 (4.98)
Alloy:	Copper-steel
MPIF specification:	FC-0205-70HT
Tensile strength, min:	480 MPa (70,000 psi)
Density:	
g/cm^3 (lb/in.^{3})	6.4-6.8 (0.23-0.25)
Relative density:	81-86%
Secondary operations:	Machining, drilling, tapping, grinding, heat treating and drawing

The cams are essential components in a variable ratio belt drive used on an agricultural combine. Two finished cams are assembled opposite each other with the cam surfaces interlocking. Relative rotation of the cams adjusts the position of the V-belt in the split pulley to increase or decrease the speed of the driven cylinder. Very close tolerances are achieved by machining the minor diameter and grinding the O.D. step and bottom surface. Holes are drilled and tapped to receive bolts.

The cyclic loading on the engaged surfaces, sliding surface wear under compressive loading, and high torque loads on the bolts require high mechanical properties. A minimum tensile strength of 480 MPa (70,000 psi) is developed by quench hardening and tempering.

CASE STUDY NO. 9

Copper Heat Sink
Process: P/M

Component name:	Heat sink
Size - mm (in.)	23 (0.91) O.D.
Weight - g (oz):	11.3 (0.4)
Alloy:	Copper
Density:	
g/cm³ (lb/in.³)	7.5 (0.271) min
Relative density:	83.6% min.
Secondary operations:	None
Alternate process:	Machined copper extrusion
Annual production:	1,000,000

High thermal conductivity, ductility and a complex configuration were the key product requirements for the automotive heat sink. Three copper heat sinks are soldered to a diode plate in an automotive alternator.

The alternate process was machining from a copper extrusion. However, the net shape P/M components cost less to produce and were more consistent.

CASE STUDY NO. 10

Self-Adjusting Mechanically Applied Brake Actuator System
Process: P/M

Component name:	Stators & Rotors	End Cover	Clutch Housing
Weight kg (lb):	0.44 (0.96)	1.05 (2.32)	0.0135 (0.30)
Alloy:	Iron-carbon	Iron-carbon	Iron-carbon
MPIF specification:	FX-1008-110HT*	FX-1008-110HT*	F-0008-30
Strength - MPa (psi)			
Tensile:	760 (110,000)	760 (110,000)	295 (42,000)
Yield:	—	—	240 (35,000)
Density:			
g/cm³ (lb/in.³)	7.5 (0.27)	7.6 (0.275)	6.9 (0.25)
% Dense	95	96	88
Secondary operations:	Machined and Heat treated	Machined and Heat treated	
Number of parts:	2**	1	1
Alternate process:	Machining	Machining	Machining
Annual production:	120,000	30,000	30,000

 * Copper infiltrated

** Includes right hand and left hand rotor and stator. One rotor and one stator used in each system.

The brake actuator system operates in commercial windmills, large articulating four-wheel agricultural tractors, large underground boring equipment, and the space shuttle transport pad, which uses approximately forty brake units. The P/M components are interchangeable among three brake systems, with four used in each system.

The actuator features a one-way roller clutch and torque-limiting slip ring, which compensate for brake pad wear. The rotor has three 13° helix angle ball pocket ramps. As the brake engagement turns the rotor, the axial displacement of the 15.8 mm (0.625 in.) balls riding up the helix ramp forces the adjusting shaft against the brake pad as a single unit.

CASE STUDY NO. 11

Electrical Connector for Surgical Instrument
Process: MIM

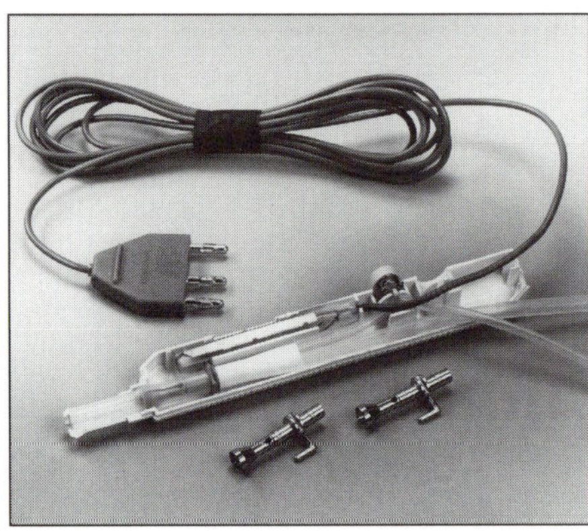

Component name:	Electrical connector
Length - mm (in.):	210 (8.25)
Alloy:	316L stainless steel
MPIF specification:	MIM 316L
Tensile strength, minimum:	480 MPa (70,000 psi)
Yield strength, minimum:	170 MPa (25,000 psi)
Hardness, minimum:	67 HRB
Elongation, minimum:	40%
Density:	
g/cm³ (lb/in.³)	7.64 (0.276)
Relative density:	96.8% (based on 7.89)
Alternate process:	Machined from bar stock

With the increasing awareness of AIDS and other blood diseases, suppliers are interested in decreasing the cost of products that are single-use disposable items. In this application, MIM allowed engineers to design a complex component capable of performing several functions at a price that justifies its being disposable.

The electrical connector is a part of a disposable handswitch that is used with an endoscope in performing abdominal surgery. The connector provides an electrical path for the surgical cutting tool. It also allows irrigation in the cutting tool to clean areas for visibility, and it provides a sealed connection to the cutting tool and hoses for suction of blood, fluids and tissue.

Dimensional precision of the connecter is essential for alignment. The diameter of the arm is held to 2.34 ± 0.025 mm (0.092 ± 0.001 in.), and its location is held to similar close tolerance.

The connector was originally machined from bar stock, and the arm was inserted into the body. Assembly problems occurred, causing the parts to be rejected. The cost of the machined arm was eight times that of the MIM production cost.

CASE STUDY NO. 12

Crimping and Cutting Tool Components
Process: MIM

Component name:	Interchangeable blades	Cam/spring receptacle
Weight - g (oz):	20 (0.7)	
Alloy:	MIM 17-4 PH Stainless Steel	Nickel-iron
Tensile strength:	965 MPa (140,000 psi)	
Hardness:	55 HRC	58-60 HRC
Density:		
g/cm^3 ($lb/in.^3$)	7.65 (0.276)	7.80 (0.282)
Relative density, %:	98.4	99.0
Secondary operations:	Coin and grind	None
Number of parts:	3 per tool	
Heat treatment:	Carbonitride and quench	Carbonitride and quench
Surface treatment:	None	Black oxide

Three interchangeable blades and a cam/spring receptacle are used in a crimping and cutting tool for telephone terminal wire connections. Metal injection molding was chosen to achieve the required material properties and dimensional precision.

On the blades, the interface features are held to ±0.050 mm (0.002 in.) to ensure a fit with molded terminal blocks. The barrel diameter is held to ±0.025 mm (0.001 in.) to insure a proper fit with the receptacle. Coining brings the blades to the desired tolerance, and grinding produces a sharp knife edge in the wire cutting plane. On the receptacle, the black oxide finish improves corrosion resistance and appearance.

CASE STUDY NO. 13

Drive Hub for Office Machines
Process: P/M

Component name:	Drive hub
Size - mm (in.):	
O.D.	12.7 (0.50)
Length	25.4 (1.00)
Weight - g (oz):	13.0 (0.45)
Alloy:	Nickel molybdenum steel
Tensile strength:	515 MPa (75,000 psi)
Yield strength:	450 MPa (65,000 psi)
Hardness:	80 HRB
Elongation:	3%
Density:	
g/cm^3 (lb/in.^{3})	6.8 (0.25)
Relative density	86.6% (based on 7.85)
Secondary operations:	Deburring
Alternate process:	Machining from steel
Annual production:	30,000

The hub is used in the drive systems of office mail inserting machines. Typically ten to twelve hubs are used in each machine. The multi-functional hub presses into pulleys and drive gears made from sheet steel, plastic, aluminum and other P/M components.

The hub was originally machined from tool steel to achieve the required tolerances and mechanical properties. The D-hole in the knurled section requires a tolerance of ±0.025 mm (0.001 in.) and a concentricity of 0.050 mm (0.002 in.) T.I.R. All dimensional tolerances and mechanical properties are being met by the P/M hub at a substantial cost reduction.

CASE STUDY NO. 14

Brush Box for Permanent Magnet Motors
Process: P/M

Component name:	Brush box
Weight - g (oz):	22 (0.78)
Alloy:	601AB Aluminum
Tensile strength, min:	144 MPa (21,000 psi)
Yield strength:	96 MPa (13,700 psi)
Elongation:	6%
Density:	
$\quad$ g/cm^3 (lb/in.^{3})	2.5 (0.090)
$\quad$ Relative density:	93%
Secondary operations:	Sizing and drilling
Surface treatment:	Clear anodizing
Alternate process:	Extrusion
Annual production:	60,000

The brush box retains a carbon brush in permanent magnet motors. The motors are used in industrial applications as well as tread mills and other exercise equipment. After compacting, sintering and sizing, several finish machining operations are performed. Various combinations of mounting holes sizes and locations allow the P/M parts producer to make four different boxes from one basic P/M form. A clear anodize treatment provides for the application of various colors specified by the end user.

The boxes had been previously made from extrusions; however, burrs that formed during the cutoff operation made the process unacceptable. P/M also produced a quieter box due to superior damping properties. Die casting was not feasible because the silicon content of die casting alloys precludes clear anodizing, and the required sound damping would not have been achieved.

The brush box is also made in a larger size, with similar shape, weighing 36 g (1.25 oz).

CASE STUDY NO. 15

Electrical Contact Pivot
Process: P/M

Component name:	Electrical contact pivot
Size - mm (in.):	41.2 (1.62) x 33.2 (1.31) x 15.9 (0.625)
Weight - g (lb):	75 (0.17)
Alloy:	99.9% Copper
Tensile strength:	255 MPa (37,000 psi)
Elongation:	18%
Density:	
g/cm³ (lb/in.³)	8.4 (0.304)
Relative density:	93.7%
Electrical conductivity	92.5% IACS at 20°C (68°F)
Secondary Operations:	Resin impregnation, cross hole drilling, coining and deburring
Surface treatment:	Silver plating
Alternate process:	Machining from plate and bar stock
Annual production:	100,000

The pivot serves as a contact in an electrical disconnect box for commercial and industrial applications. Pure copper was chosen to achieve the required electrical conductivity of 92.5% IACS (92.5% that of wrought pure copper.) This high level of conductivity required a correspondingly high density. The contact must also meet the tensile strength requirements because the disconnect is subject to high shock loading, which results from spring tension in the mechanism.

The pivots are made and brought to near net shape with a sizing operation. Only one finish machining operation, cross drilling a hole, is necessary. A silver plate is applied to improve conductivity at the contact surfaces. Plating is preceded by resin impregnation, which seals the very small amount of microporosity to prevent entrapment of plating solutions.

Prototype pivots were produced in production tools, and tested for performance and response in humidity chambers. The tests allowed the P/M parts producer to fine tune process parameters and ensure that performance standards are maintained throughout the production life of the product.

Close cooperation between the P/M part producer and the user made possible the development of a component that met service requirements which had not previously been met in commercial production. It paved the way for numerous other components used in similar applications.

CASE STUDY NO. 16

High Strength Brass Yoke for Sprinkler System
Process: P/M

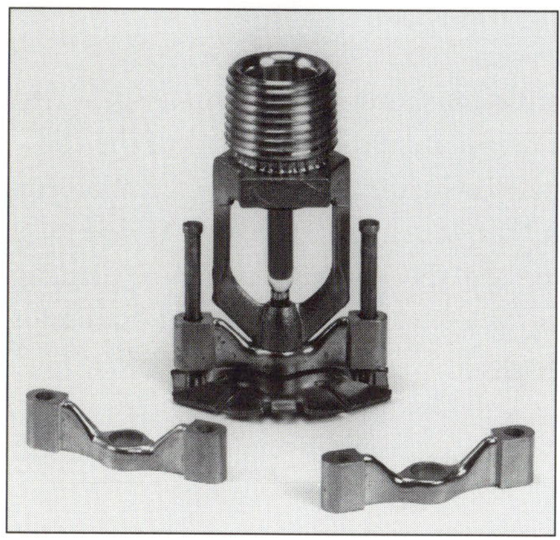

Component name:	High Strength Yoke for Sprinkler System
Length - mm (in.):	40 (1.6)
Weight - g (oz):	10.5 (0.37)
Alloy:	High strength brass
Tensile strength, typical:	275 MPa (40,000 psi)
Yield strength, minimum:	138 MPa (20,000 psi)
Density:	
g/cm^3 ($lb/in.^3$)	8.0 (0.289)
Relative density:	96.8% (Based on 8.26)
Secondary operations:	Coining and polish-deburring
Impregnation:	Resin impregnation
Annual production:	150,000

The brass yoke performs vital functions during each of two stages when the fire protection sprinkler system is activated. The first occurs when the temperature of the unit reaches 57°C (135°F) from a fire. The solder on the cover melts, releasing it to the floor, and allowing the deflector arms to drop to the working position. The yoke guides the arms, and positions them for sprinkling.

The second function occurs when the temperature reaches 68°C (155°F). At this point, a liquid filled glass bulb breaks, allowing a stream of water at pressures up to 1.2 MPa (175 psi) to flow. The contour of the yoke and its position in relation to the deflector create the required hemispherical water distribution pattern.

Very close dimensional precision is required. The 33.35 mm (1.313 in.) center line hole-to-hole dimension is held to ±0.075 mm (±0.003 in.). The 1.02/1.78 mm (0.040/0.070 in.) radii must blend very smoothly with 0.64/1.27 mm (0.025/0.050 in.) radii on two surfaces to create the smooth, dense surface required for proper deflection of the water.

The yoke is also required to pass rigorous tests to verify strength and corrosion resistance and ensure precise performance. Corrosion tests include salt spray, ammonia, carbon dioxide and sulfur dioxide. Extended intervals of test under water pressures of 1.4 MPa (200 psi) provide a safety margin against metal failure. Yoke assemblies are also tested extensively, literally "under fire".

Glossary

Absolute pore size—The maximum pore opening of a porous material, such as a filter, through which no larger particle will pass. Synonymous with maximum pore size.

Acicular powder—Needle shaped particles.

Activated sintering—A process by which the rate of sintering is significantly increased by means other than changing time or temperature; e.g., addition of a constituent to the powder, or the atmosphere, or thermal cycling.

Agglomerate—Several particles adhering together.

Air classification—Separating powder into particle size fractions by means of an air stream of controlled velocity.

Alloy powder—See Pre-alloy powder.

Angle of repose—The basal angle of a pile formed by a powder when freely poured under specified conditions onto horizontal surface.

Apparent density—The weight of a unit volume of powder, usually expressed as grams per cubic centimeter, determined by a specified method.

Apparent hardness—The value obtained by testing a sintered material with standard indentation hardness equipment. Since the reading is a composite of pores and solid material, it is usually lower than that of solid material of the same composition and condition.

Atomization—The dispersion of a molten metal into particles by a rapidly moving gas or liquid stream or by mechanical means.

Atomized metal powder—Metal powder produced by the dispersion of a molten metal by a rapidly moving gas or liquid stream, or by mechanical means.

Average pore size—The average pore diameter of a porous material, such as a filter, which conforms to a specific particle removal requirement.

Binder—A cementing medium. Either a material added to the powder to increase the green strength of the compact, which is expelled during sintering, or a material (usually of relatively lower melting point) added to a powder mixture for the specific purpose of cementing together powder particles which alone would not sinter into a strong body.

Binder removal—The chemical or thermal extraction of binder from a compact. Synonymous with Debinding.

Blank—A pressed and sintered compact, usually in the unfinished condition, requiring cutting, machining, or some other operation to give it a final shape.

Blending—The thorough intermingling of powders of the same nominal composition. (See Mixing.)

Blind pore—A dead ended pore connected to the surface.

Bridging—The formation of arched cavities in a powder mass.

Briquet(te)—See Compact.

Bulk density—The density of a powder under non-specified conditions. For example, in a shipping container.

Burn off—That stage of a sintering cycle referring to the time and temperature necessary to remove ingredients used to assist the forming of a powder metallurgy part, such as binders or die lubricants.

Carbonitriding—Introduction of carbon and nitrogen into a solid alloy (usually ferrous) by holding above the temperature at which austenite forms.

Carbonyl powder—A metal powder prepared by the thermal decomposition of a metal carbonyl.

Carburize—Introduce carbon into a solid ferrous alloy by holding above the temperature at which austenite forms.

Cemented carbides—A solid and coherent mass made by pressing and sintering a mixture of powders of one or more metallic carbides and a much smaller amount of a metal, such as cobalt, to serve as a binder.

Chemically precipitated metal powder—Powder produced by the replacement of one metal from a solution of its salts by the addition of another element higher in the electrochemical series, or by other reducing agent.

Classification—Separation of a powder into fractions according to particle size.

Closed pore—A pore not communicating with the surface.

Coining—Pressing a sintered compact to obtain a definite surface configuration (not to be confused with repressing or sizing).

Cold pressing—Forming a compact at room temperature.

Cold welding—Cohesion between two surfaces of metal, generally under the influence of externally applied pressure, at room temperature.

Comminuted powder—A powder produced by mechanical disintegration of solid metal.

Communicating pores—See Interconnected porosity.

Compact—An object produced by the compression of metal powder, generally while confined in a die, with or without the inclusion of nonmetallic constituents. Synonymous with Briquet.

Compactibility—The capacity of a metal powder to be pressed into a shape that maintains its integrity during subsequent processing. Compactibility is a function of both compressibility and green strength.

Compacting tool set—The entire set of tooling members: punches, die, core rods.

Composite compact—A metal powder compact consisting of two or more adhering layers, rings, or other shapes of different metals or alloys with each layer retaining its original identity.

Composite powder—A powder in which each particle consists of two or more separate materials.

Compound compact—A metal powder compact consisting of mixed metals, the particles of which are joined by pressing or sintering or both, with each metal particle retaining substantially its original composition.

Compressibility—The capacity of a metal powder to be compacted uniaxially in a closed die. Compressibility may be expressed numerically as the pressure to reach a required density or, alternately, the density at a given pressure.

Continuous sintering—Presintering, or sintering, in such manner that the objects are advanced through the furnace at a fixed rate by manual or mechanical means.

Core, coring—(MIM) Non-functional blind or through hole used to reduce weight and improve wall uniformity in MIM.

Core rod—The separate member of the compacting tool set that forms a hole in the component.

Creep—The time-dependent strain occurring under stress at high temperatures.

Cut—See Fraction.

Debinding—See Binder removal.

Dendritic powder—Particles, usually of electrolytic origin, having the typical pine tree structure.

Density (dry)—The mass per unit volume of an unimpregnated P/M part.

Density (wet)—The mass per unit volume of a P/M part impregnated with oil or other nonmetallic materials.

Density ratio—The ratio of the determined density of a compact to the absolute density of metal of the same composition, usually expressed as a percentage.

Die—The part or parts making up the confining form

in which a powder is pressed. Synonymous with Mold.

Die body—The stationary or fixed part of a die.

Die insert—A removable liner for the die assembly. Synonymous with die liner.

Die lubricant—A lubricant mixed with the powder or applied to the walls of the die and punches to facilitate the pressing and ejection of the compact.

Die set—A part of a press that holds and aligns the tooling in proper relation to the press.

Dimensional change—See Growth; see Shrinkage.

Dimensional control—High pressure compaction and controlled sintering to ensure uniform compact shrinkage.

Dispersion-strengthened material—A material consisting of a metal and finely dispersed, substantially insoluble, metallic or nonmetallic phase.

Dissociated ammonia—A reducing gas produced by the thermal decomposition of anhydrous ammonia over a catalyst, resulting in a gas of 75% hydrogen and 25% nitrogen.

Double-action pressing—A method by which a powder is pressed by relative motion between two of the three prime tooling members (upper punch, lower punch and die).

Draft—(MIM) Taper on walls to facilitate removal of a component from the mold.

Ejector pin—(MIM) Tool member used in MIM to push the component from the mold.

Electrolytic powder—Powder produced by electrolytic deposition or the pulverization of an electrodeposit.

Elemental powder—Powder of a single chemical species like iron, nickel, titanium, copper or cobalt, with no alloying ingredients.

Endothermic atmosphere (gas)—A reducing gas atmosphere used in sintering, produced by the reaction of a hydrocarbon fuel gas and air over a catalyst with the aid of an external heat source. The resulting atmosphere is low in carbon dioxide and water vapor with relatively large percentages of hydrogen and carbon monoxide. Maximum combustibles approximately 60%.

Equi-axed powder—See Granular powder.

Eutectic—An alloy having the composition indicated by the eutectic point on an equilibrium phase diagram.

Exothermic atmosphere (gas)—A reducing gas atmosphere used in sintering, produced by partial or complete combustion of a hydrocarbon fuel gas and air. Maximum combustibles approximately 25%.

Exudation—The action by which all or a portion of the low melting constituent of a compact is forced to the surface during sintering. Sometimes referred to as "bleed out." Synonymous with Sweating.

Feedstock—(MIM) A moldable mixture of metal powder and binder.

Ferrite—In the field of magnetics, substances having the general formula: M^+O, $M_2^{+++}O_3$ the trivalent metal often being iron.

Fill ratio—The ratio of the height of the loose powder to the height of the compact made from it.

Fines—The portion of a powder composed of particles whose thickness is smaller than a specified size, currently less than 44 micrometres. See also Superfines.

Flake powder—Flat or scale-like particles whose thickness is small compared with the other dimensions.

Flash, flashing—Small burr(s) caused by feedstock or preform penetrating between mating tool members.

Flow rate—The time required for a powder sample of standard weight to flow through an orifice in a standard instrument according to a specified procedure.

Fluid permeability—See Permeability (fluid).

Forging Preform—(P/F) An unsintered, presintered, or fully sintered compact intended to be subjected to hot or cold forming and densification.

Fraction—That portion of a powder sample which lies

between two stated particle sizes. Synonymous with Cut.

Gas classification—The separation of powder into particle size fractions by means of a gas stream of controlled velocity.

Gate—(MIM) The opening through which feedstock flows to enter the mold cavity in MIM.

Granular powder—Particles having approximately equidimensional, non-spherical shapes.

Granulation—A general term for the production of coarse metal powders using methods such as: a. Pouring molten metal through low pressure water jets, b. Violent agitation of the molten metal while solidifying, c. Agglomeration of smaller particles.

Green—Unsintered (not sintered); for example: green compact, green density, green strength.

Green density—The density after pressing a powder prior to sintering.

Green expansion—The increase in dimensions of a compact relative to the die dimensions after being ejected from the die.

Green strength—The strength of the as-pressed powder compact.

Growth—An increase in dimensions of a compact occurring during sintering (converse of shrinkage).

Hot densification—The rapid deformation of a heated powder preform in a die set, primarily in the pressing direction, in order to reduce porosity. Synonymous with Hot repressing.

Hot forging—See P/F.

Hot pressing—The simultaneous heating and molding of a compact.

Hot repressing—See Hot densification.

Hot upset powder forging—(P/F) Hot densification of a P/M preform by forging where there is a significant amount of material flow.

Hot repress powder forging—(P/F) Hot densification of a P/M preform by forging where material flow is mainly in the direction of pressing. Also referred to as *hot restriking* or *hot coining.*

Hot working—The plastic deformation of a metal at a temperature and strain rate that minimizes work hardening.

Hydrogen loss—The loss in weight of metal powder or of a compact caused by heating a representative sample for a specified time and temperature in a purified hydrogen atmosphere-broadly a measure of the oxygen content of the sample, when applied to materials containing only such oxides as are reducible with hydrogen and no hydride forming material.

Hydrogen reduced powder—Powder produced by the hydrogen reduction of a metal oxide.

Impregnation—A process of filling the pores of a sintered compact with a nonmetallic material such as oil, wax or resin.

Infiltration—A process of filling the pores of a sintered, or unsintered, P/M compact with a metal or alloy of lower melting point.

Infiltration efficiency—The ratio in percent of the amount of infiltrant absorbed by the component to the amount of infiltrant originally used. See MPIF standard 49.

Infiltration erosion—The solution of the base metal at the contact area between the component and the molten infiltrant where the infiltrant flows into the component causing pits, channels and coarse surface porosity. See MPIF standard 49.

Infiltration loading density—Infiltrant mass per area of contact between infiltrant and component. See MPIF standard 49.

Infiltration residue—Material that remains on the surface of the component after infiltration. See MPIF standard 49.

Injection molding—See Metal injection molding.

Interconnected porosity—A network of contiguous

pores in and extending to the surface of a sintered compact. Usually applied to P/M materials where the interconnected porosity is determined by impregnating the specimens with oil.

Interparticle friction—The friction associated with fine powders which limits sliding, packing and densification.

Irregular powder—Powder having particles lacking symmetry.

Isostatic pressing—Pressing a powder by subjecting it to nominally equal pressure from every direction.

K-factor—The strength constant in the formula for "radial crushing strength" of a plain sleeve specimen of sintered metal. See Radial crushing strength.

Knock-out pin—(MIM) See Ejector pin.

Lamination—A series of planar cracks within the pressed compact typically normal to the direction of pressing. Synonymous with pressing crack and slip crack.

Liquid phase sintering—Sintering a P/M compact, or loose powder aggregate, under conditions where a liquid phase is present during part of the sintering cycle.

Loading—Filling a die cavity with powder.

Lower punch—The member of, the compacting tool set that determines the volume of powder fill and forms the bottom of the component being produced.

Lubricant—An agent mixed with or incorporated in a powder to facilitate the pressing and ejecting of the compact.

Magnetic permeability—See Permeability (magnetic).

Master alloy powder—A pre-alloyed powder of high concentration of alloy content designed to be diluted when blended with a base powder to produce the desired composition.

Matrix metal—The continuous phase of polyphase alloy or mechanical mixture; the physically continuous metallic constituent in which separate particles of another constituent are embedded.

Maximum pore size—See Absolute pore size.

Mechanical alloying—Forming an alloy powder by milling elemental powders for a prolonged time. Frequently used to create dispersion strengthened alloy powders.

Mechanical component—A shaped body pressed from metal powder and sintered wherein self-lubrication is not a primary property. A component but not an oil-impregnated bearing. See Powder metallurgy part.

Mesh—The screen number of the finest screen through which substantially all of the particles of a given sample will pass. The number of screen openings per linear inch of screen.

Metal filter—A metal structure having controlled interconnected porosity produced to meet filtration or permeability requirements.

Metal injection molding—See MIM.

Metal powder—Discrete particles of elemental metals or alloys normally within the size range of 0.1 to 1000 micrometers.

Milling—The mechanical treatment of metal powder, or metal powder mixtures, as in a ball mill, to alter the size or shape of the individual particles, or to coat one component of the mixture with another.

MIM—The acronym representing Metal Injection Molding. A process similar to plastic injection molding that uses a metal powder mixed with a binder. Generally, this is followed by debinding and sintering. The process is used to form components with intricate shapes that cannot be formed by conventional P/M compaction processes.

MIMA—The Metal Injection Molding Association. One of the trade associations that constitutes the Metal Powder Industries Federation (MPIF).

Minus sieve—The portion of a powder sample which passes through a standard sieve of specified number. See Plus sieve.

Mixed powder—A powder made by mixing two or more powders as uniformly as possible. The constituent powders will differ in chemical composition and/or in particle size and/or shape.

Mixing—The thorough intermingling of powders of two or more different compositions

Mold—See Die.

Molding—Pressing powder to form a compact.

Molding press—A press used to form compacts from powder.

Monodisperse—All particles are the same size.

Multiple pressing—A method of pressing whereby two or more compacts are produced simultaneously in separate die cavities.

Multiple action pressing—Pressing action that utilizes several members so arranged and controlled that a separate member supports each level of a multilevel component. The action assures minimum density gradients in all levels of the component.

Near net shape—A compact that has the general shape of the final product, but requires machining of threads and through holes perpendicular to the direction of pressing.

Neck formation—The development of a necklike bond between particles during sintering.

Net shape—A compact fabricated to final density and dimensions.

Nodular powder—Irregular particles having knotted, rounded, or similar shapes.

Oil content—The measured amount of oil contained in an oil-impregnated object. For example a self-lubricating bearing.

Open pore—A pore communicating with the surface.

Oversize powder—Particles coarser than the maximum permitted by a given particle size specification.

Partially alloyed powder—A metallic powder composed of two or more elements that are partially alloyed in the powder manufacturing process.

Particle size—The controlling lineal dimension of an individual particle as determined by analysis with sieves or other suitable means.

Particle size distribution—The percentage by weight, or by number of each fraction, into which a powder sample has been classified with respect to sieve number or micrometres. (Preferred usage: "particle size distribution by weight" or "particle size distribution by frequency.")

Parting line—(MIM, P/F) Witness mark at the interface of mating tool members.

Permeability (fluid)—The rate of fluid flow through a porous material under specified conditions of area, thickness and pressure.

Permeability (magnetic)—A measure of the ease of a material to be magnetized. Not to be confused with fluid permeability.

P/F—The acronym representing powder forging. A process in which unsintered, presintered, or sintered metal preforms are hot formed in confined dies. Sometimes called P/M forging or P/M hot forming. There are two basic forms: Hot upset powder forging and hot repress powder forging.

Platelet powder—Flat particles of metal powder having considerable thickness (as compared to flake powder).

Plus sieve—The portion of a powder sample retained on a standard sieve of specified number. See Minus sieve.

P/M—The acronym representing powder metallurgy. Used as P/M part, P/M product, P/M process, etc.

P/M part—A shaped object that has been formed from metal powders and bonded by heating below the melting point of the major constituent. A structural or mechanical component, made by the powder metallurgy process.

PMPA—The Powder Metallurgy Parts Association. One of the trade associations that constitutes the Metal Powder Industries Federation (MPIF).

Pore—An inherent or induced cavity within a particle or within an object.

Pore forming material—A substance included in a powder mixture that volatilizes during sintering and thereby produces a desired kind and degree of porosity in the finished compact.

Porosity—The amount of pores (voids) expressed as a percentage of the total volume of the powder metallurgy part.

Pore size—The average pore diameter of a porous material such as a filter, which conforms to specific particle removal requirements, usually the removal of 95% to 100% of a given particle size distribution.

Porous metal—A metal structure having controlled interconnected porosity. See Metal filter.

Powder—Particles of matter characterized by a small size, less than 1 mm in size.

Powder flow meter—An instrument for measuring the rate of flow of a powder according to a specified procedure.

Powder forging—See P/F.

Powder metallurgy—The processes of producing metal powders and their consolidation into shaped objects.

Powder metallurgy part—See P/M part.

Powder rolling—The progressive compacting of metal powders by the use of a rolling mill. Synonymous with Roll compacting.

Pre-alloyed powder—A metallic powder composed of two or more elements which are alloyed in the powder manufacturing process, and in which the particles are of the same nominal composition throughout. Synonymous with Alloy powder.

Preform—(P/F) The initially pressed and sintered compact to be subject to forging.

Preforming—The initial pressing of a metal powder to form a compact which is subjected to a subsequent pressing operation other than coining or sizing. Also, the preliminary shaping of a refractory metal compact after presintering and before the final sintering.

Premix—A uniform mixture of ingredients to a prescribed analysis, prepared by the powder producer, for direct use in compacting powder metallurgy products.

Presintering—Heating a compact at a temperature, usually to increase the ease of handling or shaping the compact, or to remove a lubricant or binder prior to sintering.

Pressed bar—A compact in the form of a bar; a green compact.

Pressed density—The mass per unit volume of an unsintered compact. Synonymous with Green density.

Pressing crack—See Lamination.

Pulverization—The reduction in size of metal powder by mechanical means.

Punch—Part of a die or compacting tool set that is used to transmit pressure to the powder in the die cavity. See upper punch and lower punch.

Radial crushing strength—The relative capacity of a plain sleeve specimen made by powder metallurgy to resist fracture induced by a force applied between flat parallel plates in a direction perpendicular to the axis of the specimen.

Rapid solidification—Production of powders from the molten state at a cooling rate above 10,000°C/ second.

Rate-of-oil-flow—The rate at which a specified oil will pass through sintered porous compact under specified test conditions.

Reduced metal powder—Metal powder produced, without melting, by the chemical reduction of metal oxides or other compounds.

Restriking—The application of pressure to a previously pressed and sintered compact, usually for the purpose of improving physical or mechanical properties and dimensional characteristics.

Rib—(MIM, P/F) Thin wall added to a component in MIM and P/F to increase strength or rigidity but having no other function. In P/F, a rib is oriented in the direction of forging.

Roll compacting—See Powder rolling.

Rolled compact—A compact made by passing metal powder continuously through a rolling mill so as to form relatively long sheets of pressed material.

Rotary press—A machine fitted with a rotating table carrying multiple dies in which a material is pressed.

Runner—(MIM) A portion of the feed path in the mold base used to transport feedstock to the cavity in MIM.

Screen analysis—See Sieve analysis.

Segment die—A die contour that is fabricated by assembling individual die inserts which are retained by a case or ring.

Segregation—The undesirable separation of one or more components of a powder.

Semi-alloyed powder—See Partially alloyed powder.

Shrinkage—A decrease in dimensions of a compact which occurs during sintering (converse of growth).

Sieve analysis—Particle size distribution; usually expressed as the weight percentage retained upon each of a series of standard sieves of decreasing size and the percentage passed by the sieve of finest size. Synonymous with Screen analysis.

Sieve classification—The separation of powder into particle size ranges by the use of a series of graded sieves.

Sieve fraction—That portion of a powder sample which passes through a standard sieve of specified number and is retained by some finer sieve of specified number.

Single-action pressing—A method by which a powder is pressed by motion of one of the two punches.

Sink marks—(MIM) Visible depressions in a molded or sintered MIM component caused by localized shrinkage.

Sintering—Metallurgically bonding particles in a powder mass or compact resulting form thermal treatment at a temperature below the melting point of the main constituent.

Sizing—A final pressing of a sintered compact to secure a desired size of a dimension.

Slip casting—A method of forming metal shapes by pouring a stabilized water-suspension of metal powders into the shaped cavity of a fluid absorbing mold, diffusing the liquid into the mold wall, removing the casting from the mold and sintering.

Slip crack—See Lamination.

Solid-state sintering—Sintering of powder or compact without formation of a liquid phase.

Specific surface—The total surface area of the powder particles per unit mass of powder.

Spherical powder—Powder consisting of round or globular shaped particles.

Split die—A two-piece die that splits in a horizontal plane to allow ejection of components that could not be ejected from a one-piece die.

Sponge iron powder—Iron powder of porous structure usually produced by the reduction of iron oxide.

Spongy—A porous condition in metal powder particles usually observed in reduced oxides.

Spring back—Elastic expansion upon release of the compacting or coining forces.

Stripper punch—A punch which, in addition to forming the top or bottom of the die cavity, later moves further into the die to eject the compact.

Subsieve fraction—Particles, all of which will pass through a 44 micrometre (no. 325) standard sieve.

Superfines—The portion of a powder composed of particles that are smaller than a specified size, currently less than 10 micrometres.

Sweating—See Exudation.

Swelling—The growth in compact dimensions due to pore formation. A low melting point additive or an entrapped gas can form pores during sintering and thereby give swelling.

Tap density—The apparent density of a powder obtained when the receptacle is tapped or vibrated during loading under specified conditions.

Thermoplastic—(MIM) Organic material that becomes plastic each time it is heated as opposed to a thermoset, which can not be returned to the plastic state after heating.

Transverse rupture strength—The stress, calculated from the flexure formula, required to break a specimen as a simple beam supported near the ends and applying the load midway between the fixed center lines of the supports.

Upper punch—The member of the compacting tool set that forms the top of the component.

Walking Beam Furnace—A continuous sintering furnace that transports containers or trays of P/M components by lifting and moving them in steps.

Warpage—Distortion which may occur in a compact during sintering.

Web—(MIM, P/F) Thin wall added to a component in MIM and P/F to increase strength or rigidity but having no other function. In P/F, a web is oriented perpendicular to the direction of forging.

Withdrawal process—An operation by which the die descends over a fixed lower punch to achieve ejection of the compact.

Bibliography

Bolze, G.A. and Capus, J.M., Factors Affecting Magnetic Properties of Sintered Iron, Modern Developments in Powder Metallurgy Vol 11, Hausner, H.H. and Taubenblat, P.W., Ed., Metal Powder Industries Federation, Princeton, NJ, 1977, pp 355-370.

German, R.M., Powder Metallurgy Science, Second Edition, Metal Powder Industries Federation, Princeton, NJ, 1994.

ASTM Standard A596 "Standard Method of Test for D.C. Magnetic Properties of Material Using Ring Test Procedures and the Ballistic Method", American Society for Testing and Materials, Philadelphia, PA.

Moyer, K.M. The Magnetic Properties (DC) of Atomized Iron Powder Cores, Modern Developments in Powder Metallurgy Vol 11, Hausner, H.H. and Taubenblat, P.W. Ed., Metal Powder Industries Federation, Princeton, NJ, 1977, pp 371-384.

Walker, E.V. Worn, D.K. and Walters, R.E.S., "Application of Powder Metallurgy to the Production of High Permeability 69 Magnetic Alloy Strip", Symposium on Powder Metallurgy, Special Report No. 58, Iron and Steel Institute, London, 1956, pp 204-208.

Jones, W.D., Fundamental Principles of Powder Metallurgy, London, 1960, pp 680-710.

Schumacher, E., Magnetic Powders and Production of Cores for Induction Coils, Powder Metallurgy, Wulff, J., Ed., American Society for Metals, 1942, pp 166-172.

Kulkarni, K.K. ed., New Perspectives in Powder Metallurgy Vol. 8, Powder Metallurgy for Fully Dense Products, Metal Powder Industries Federation, Princeton, NJ, 1987.

Hausner, H.H., Johnson, P.K., and Roll, K.H., Ed, New Perspectives in Powder Metallurgy, Volume Six, Forging of Powder Metallurgy Preforms, Metal Powder Industries Federation, Princeton, NJ, 1973.

Lenel, F.V., Powder Metallurgy Principles and Applications, Metal Powder Industries Federation, Princeton, NJ, 1980.

Nayar, H.S., Production Sintering Atmospheres, Metals Handbook, 9th Ed, Vol. 7, American Society for Metals, Metals Park, OH, 1984.

High Temperature Sintering, Sanderow, H.I., Ed., Metal Powder Industries Federation, Princeton, NJ, 1990.

Kuhn, H.A. and Ferguson, B.L. Powder Forging, Metal Powder Industries Federation, Princeton, NJ, 1990.

Material Standards for P/M Structural Parts, MPIF Standard 35, 1994 edition, Metal Powder Industries Federation, Princeton, NJ.

Material Standards for Metal Injection Molded Parts, 1993-1994 edition, Metal Powder Industries Federation, Princeton, NJ.

Material Standards for P/M Self-Lubricating Bearings, 1991-1992 edition, Metal Powder Industries Federation, Princeton, NJ.

Lall, C., Soft Magnetism, Fundamentals of Powder Metallurgy and Metal Injection Molding, Monographs in P/M Series No. 2, Metal Powder Industries Federation, Princeton, NJ, 1992.

Lawley, A., Atomization, The Production of Metal Powders, Monographs in P/M Series No. 1, Metal Powder Industries Federation, Princeton, NJ, 1992.

Standard Test Methods for Metal Powders and Powder Metallurgy Products, 1995 edition, Metal Powder Industries Federation, Princeton, NJ.

Index

A

Abrasion resistance 11
Accidental or unanticipated conditions 11, 12
Alloy designation code 70
    copper alloys 74
    ferrous alloys 70
    P/F 73, 79
Alloying 67, 69, 93, 96
    nitrogen 74
Alloys
    aluminum 75
    bearing 56, 76
    composition and properties 67
    copper 74
    ferrous 70
    magnetic 72
    MIM 40, 77, 107
    P/F 22, 77
    P/M 1, 17, 58, 79
    stainless 73
Alphanumeric characters 26
Aluminum 1, 20, 40, 61, 75, 85
Aluminum oxide 61
Anodizing 20, 62, 75
Assemblies, one-piece 1, 39
Bearing surfaces 33
Bearings 4
    alloys 57
    design 58
    integral 55, 58
    K-factor 54
    materials 74
    operating modes 56
    properties 70, 71, 75, 76, 80, 85
    self-lubricating 55
    specifying 66

B

Bevels 31
Blackening 17, 61, 102
Bolting 39
Bosses 9, 24, 32, 33, 45
Brazing 17, 18, 39, 97, 98, 104
    alloy 104
    during sintering 40
    furnace 104
    joint 104
Burnishing 17, 18, 59, 84, 98, 101, 106
Burrs, eliminating 17, 31, 60, 100, 101
Buyer's guides 10

C

Cams 1, 32, 36, 37
Case hardening 83, 99, 100, 107
Cementing 39
Chamfers 31, 50, 101
Chemical conversion coatings 61
Chromates 61
Classes of components 88, 90
Coining 15, 17, 22, 24, 26, 38, 98, 100
Cold isostatic pressing (CIP) 1, 87
Compaction 35, 67, 87, 97
    double action 88, 89
    multiple action 88
    P/F preform 107
    pressure 14, 15, 88, 97
    single action 88-90
Component complexity 6, 8, 12, 14, 16, 19, 22-24,
       36, 40-42, 52, 53, 62, 90, 104-106
Component size 16, 17, 22, 23, 40, 72, 88, 107
Computerized Numerical Control (CNC) 21
Copper 1, 11-13, 15, 54, 55, 74, 76, 77, 85, 89
Corners
    eliminating sharp 45, 51
    removing sharp 30, 34, 60
Corrosion 11, 20, 57, 59-61, 65, 73, 75, 76, 83, 85,
      102, 103
    atmospheric 12
    environmental 62, 63
    galvanic 12, 62, 63
Counterbores 26, 29
Countersinks 26, 32
Creep 12, 54, 55

D

Dampening 4, 75
Density 53, 54, 77, 79, 80, 100, 103, 106, 107
    apparent 88
    gradients 23-26, 48, 53, 58, 65, 81, 83, 89, 92
    green component 69
    uniform 16, 49, 62
Dent resistance 54, 55
Design drawings 10, 30, 65
Die casting 1, 6, 19
Differential expansion 39
Dimensional tolerances 23
    MIM 41, 48, 106
    P/F 6, 51, 52
    P/M 23, 58, 59, 61
Dimensions
    axial 23
    radial 23, 24, 26
Draft 1, 13
    MIM 41, 42
    P/F 51
    P/M 25, 27, 28, 32-34
Drilling 17, 32, 79, 101, 106
Ductility 6, 11, 12, 57, 72-76, 79, 83, 85, 94, 106, 112

E

Electromagnetic materials 72
Electroplating 59
    aluminum 61
Elongation 61
    in P/M materials 83, 94
    of MIM steels 77
Endurance limit 54, 77, 80, 94
Engineered alloys 11
Engineered materials 67

F

Fastening and joining 22, 39
Fatigue, improving surface 61
Fillets
    MIM 45
    P/M 30
Filters 4, 69, 73, 80, 81
    filtering characteristics 63
    specifying 66
Fine blanking 1, 9, 20
Finishing
    aluminum P/M 75
    P/M 17, 18, 101
    vibratory 60

Finite element analysis (FEA) 53, 62, 63
Fixtures, inspection 65
Flange
    P/M 38
Flanges
    P/M 28, 29, 32
Flatness 23
Form 9, 22
Foundry casting 2, 20
Function 9, 40

G

Gauging methods 66
Gears 9, 36, 53, 61, 80, 109
    bevel 36
    helical 36
    miter 36
    runout tolerances 23
    specifying 65
Graphite 56-58, 76, 77, 100, 103
Grinding 17, 18, 24, 36, 37, 84, 100, 101, 106
Grooves 21, 34
    annular 38
Growth 93, 97

H

Hardenability 77, 85, 96, 100, 104, 112
Hardening
    case 99, 100, 107
    local 99
    neutral 99
    surface 100
Hardness 79, 94, 98
Hardness (apparent) 65, 83
Heat treatment 11, 17, 22-24, 57, 61, 71, 79, 103, 107, 112
    types 99
Holes
    blind 1, 28
    cross 30, 100
    D 32, 39
    lightening 9, 16, 30
    long 1
    maximum diameter 30
    MIM 43
    minimum diameter 30
    shaped 32
    tapered 28, 30
    tapped 12, 55
    through 28

Hot isostatic pressing (HIP) 1, 87
Hubs 23, 32, 37, 108

I
Impregnating 99
Impregnation 4, 11, 17, 59, 98
    for sound dampening 80
    oil 74, 99-101
    polyester and anaerobic 59
    resin 60, 61, 99-101, 103
    with oil 56, 58
Impression die forging 22
Infiltration 4, 11, 15, 17, 36, 39, 40, 54, 58, 59, 63,
        67, 72, 97
Interfacing components and systems 11, 12
Investment casting 1, 21, 77

K
Keyways 28, 32, 33, 39, 97
Knurls 32, 48

L
Length
    studs or holes, MIM 45
    to diameter ratio, P/M 33
    to wall thickness ratio, P/M 25
Length, limits for P/M 23
Levels of features
    P/F 48
    P/M 90
Loads, applied 11, 52
    bearing test 54
Lubricants, solid 4, 56
Lubrication 4, 60

M
Machinability 18, 60, 79, 97, 99
Machining 100
    guidelines 100
    MIM 106
    of P/M bearings 58
    responsibility 65
Machining procedures
    porous and self lubricating components 100
Metal removal 17
Micro-Porosity 80
    controlled 4, 67, 69, 89
    effect on corrosion resistance 83
    effect on ductility 83
    effects on secondary operations 18

in bearing surfaces 56
in bearings 55
in gears 36
in P/F 6
interconnected 56, 58, 76
limitations on secondary operations 58, 61, 98,
        101, 103
MIM 11
P/F 11
Milling 17, 101
Modulus of elasticity 13, 53

N
Net shape 1, 98

O
Oil dipping
    after blackening 61, 102
    after plating 60
    after steam treating 61, 103
Oil holding capacity 24, 56, 58

P
Particle size 67, 69, 77, 81
    P/M vs MIM 105
Passivation 61
Permeability
    fluid 81
    magnetic 72, 73
Phosphates 61
Plating 17, 72, 98, 103
    cadmium 60
    chromium 60
    copper 59
    multiple 60
    nickel 59
    zinc 60
Pore treatment processes 17
Powders
    characteristics 67
    elemental 69, 96
    methods of manufacturing 67
    MIM 107
    partially alloyed 69
    prealloyed 69
Press size
    P/M 15
Process controls 17
Product cost
    compacting 15

materials 13
MIM debinding and sintering 16
MIM injection molding 16
P/F compacting and sintering 16
repressing or forging 17
sintering 15
tooling 14
Product objectives 10, 58
Production rates 1, 88
Profilometer readings 84
Proof test 65
Properties
    engineered 13
    engineering 67, 79
    mechanical 12, 13, 53
    mechanical, P/F 79
    mechanical, P/M 81
Prototyping 10, 62

Q
Quenching 18, 74, 99

R
Radii 26, 30, 31, 37, 43, 45, 51
    on P/F components 51
Radius 32
Ratchets 36-38
Relaxation 12
Repressing 2, 25, 31, 83, 84, 98
    for flatness 24
    for improved tolerances 24
    P/F 14, 16, 17, 107, 109
Resin impregnation 60, 61, 100, 101, 103
Riveting 39
Runout 36, 37

S
Scrap losses 20, 72
Screw machining 1, 21
Secondary operations 13, 17, 84
    MIM 18, 106
    P/F 19, 112
    P/M 17, 87, 98
Service conditions 65
Shot peening 17, 61
Shrinkage
    in MIM 43
    in P/F 52
    in P/M sintering 93, 95
    in the weld zone 103

Sinter bonding 1, 97
Sinter brazing 40
Sinter-Hardening 95
Sintering 39, 93
    atmosphere 96
    high temperature 22, 83, 93
    liquid phase 84, 95
    MIM 16, 42, 106
    mixed phase 96
    P/F 16, 109, 111
    P/M 15
Sizing 15, 17, 22, 24, 98, 102
Slots 9, 34, 97
Soldering 39, 60, 97
Specifying P/M 58, 64
Spherical shapes 25
Splines, specifying 65
Springback 89, 93
Sprockets, specifying 65
Staking 12, 39, 57, 76
Stamping 1, 14, 15, 20
Standard 35, MPIF 57, 58, 64, 70, 77, 79-81, 84, 85
Standard 42, MPIF 80
Steam treating 4, 17, 18, 59, 61, 102, 104
Steps 25, 28, 31
Strength 63, 82, 100
    bearing crushing 76
    compressive 53, 61
    fatigue 6, 19, 22, 54, 61, 69, 111
    green 69, 70, 87, 88
    in joints 39
    of bearings 57
    shear 40
    tensile 19, 53, 61
    ultimate 11
    yield 55, 69, 74
Structural criteria 52
Studs 33, 45
Surface finish 48
    MIM 48
    P/M 84
    specifying 66
Surface treatment 10, 12, 58
    mechanical methods 60
    MIM 18, 106
    P/F 19, 112
    specifying 65
    types 101
Swaging 12

T
Tapers 27
    reverse 38
Tapping 17, 32, 46, 101, 106
Temperature, effects on mechanical properties 11,
        55
Temperatures
    cyclic 11
    environmental 11
    steady state 11
Thermal expansion 12, 104
Threads 21, 39, 100, 101
    MIM 46
Tool wear
    P/F 52
    P/M 24

Tooling 2, 10, 13, 19
    MIM 14, 16
    P/F 14
    P/M 14
Turning 17, 101, 106

U
Undercuts 38, 39, 46
Upsetting 39

W
Wall thickness
    MIM 40, 43
    P/M 25
Welding 17, 39, 103
    MIM 107
    P/F 112

Notes

Notes